HOW TO SUCCEED AS A SCIENTIST
From Postdoc to Professor

This unique, practical guide for postdoctoral researchers and senior graduate students explains, stage by stage, how to gain the necessary research tools and working skills to build a career in academia and beyond. The book is based on a series of successful training workshops run by the authors, and is enriched by their extensive interdisciplinary experience as working scientists.

- Discusses the tools needed to become an independent researcher, from writing papers and grant applications, to applying for jobs and research fellowships.
- Introduces skills required as an academic, including managing and interacting with others, designing a taught course and giving a good lecture.
- Concludes with a section on managing your career, explaining how to handle stress, approach new challenges and understand the higher education system.

Packed with helpful features encouraging readers to apply the theory to their individual situation, the book is also illustrated throughout with real-world case studies that enable readers to learn from the experiences of others. It is a vital handbook for all those wanting to pursue a successful academic career in the sciences.

BARBARA GABRYS is an experimental physicist with expertise in the structure and dynamics of soft matter. She has substantial experience in exploring different science disciplines through research, teaching and learning activities. Dr Gabrys was appointed Academic Advisor for the Mathematical, Physical and Life Sciences Division at the University of Oxford in 2007. She is a Fellow of the Institute of Physics and a Fellow of the Higher Education Academy.

JANE LANGDALE is a plant biologist with over 25 years' research experience in both UK and US universities. Her main research focuses on understanding the genetic basis of plant developmental processes and elucidating how those processes evolved. Professor Langdale was appointed as a University academic in 1994 and most recently has been Head of the Department of Plant Sciences at the University of Oxford. She was elected a member of the European Molecular Biology Organization (EMBO) in 2007.

D1157862

HOW TO SUCCEED AS A SCIENTIST

From Postdoc to Professor

BARBARA J. GABRYS

Department of Materials, University of Oxford

and

JANE A. LANGDALE

Department of Plant Sciences, University of Oxford

CAMBRIDGE
UNIVERSITY PRESS

CAMBRIDGE UNIVERSITY PRESS
Cambridge, New York, Melbourne, Madrid, Cape Town,
Singapore, São Paulo, Delhi, Tokyo, Mexico City

Cambridge University Press
The Edinburgh Building, Cambridge CB2 8RU, UK

Published in the United States of America by Cambridge University Press, New York

www.cambridge.org
Information on this title: www.cambridge.org/9780521765862

First published 2012

Printed in the United Kingdom at the University Press, Cambridge

A catalogue record for this publication is available from the British Library

Library of Congress Cataloguing in Publication data
Gabrys, Barbara J.
How to succeed as a scientist : from postdoc to professor / Barbara
J. Gabrys and Jane A. Langdale.
p. cm.
Includes bibliographical references and index.
ISBN 978-0-521-76586-2 (hardback)
1. Science – Vocational guidance. I. Langdale, Jane A. II. Title.
Q147.G33 2011
502.3–dc23
2011030687

ISBN 978-0-521-76586-2 Hardback
ISBN 978-0-521-18683-4 Paperback

To our parents
For life, love, sacrifices and shared wisdom.
And for never allowing us to think that being female could,
or would, hinder our career aspirations

Contents

Preface

This book is based on a series of 20 workshops developed by Jane Langdale in 2005 for postdocs in the Department of Plant Sciences, University of Oxford. The topics were subsequently extended by Barbara Gabrys to cover other disciplines in the Mathematical, Physical and Life Sciences Division at Oxford. The motivation for the workshops and for the book, stemmed from a desire to help postdocs gain a thorough understanding of what being a successful academic entails, and to provide a set of tools to help them achieve that goal. The book can also act as a foundation for others who wish to run their own series of workshops – in each chapter we give an example of how we cover the topic.

We have written the book primarily in the context of the UK higher education sector. However, much of the content is equally applicable elsewhere. The main differences relate to the titles of the various academic jobs in different countries rather than to the expectations of what those jobs entail. Specifically – 'Lecturer' in the UK is equivalent to 'Assistant Professor' elsewhere; 'probation' is equivalent to 'tenure-track'; and 'Head of Department' is equivalent to 'Chair of Department' (although Heads normally line manage academic staff whereas Chairs do not). Wherever possible we have used the more universal term Principal Investigator (PI) in order to avoid confusion.

Of course it is inevitable that everyone has to find their own path to success and has to develop their own way of doing things. What is presented in this book is very much a personal view based on our own experiences, it is certainly not meant to be prescriptive. Some of the suggestions will appeal to you and others may not – but it is always easier to modify an approach than to start from scratch. The bottom line is that most successful scientists thrive and excel because they are passionate about their subject – 'work–life' balance is essentially a misnomer because science is such an integral part of who they are. But there are skills that have to be mastered and the journey from postdoc to professor can be a challenging one. All too often, however, people focus entirely on the endpoint and forget what an exciting and fulfilling career we have – our main piece of advice is – 'don't forget to enjoy the journey'.

Barbara J. Gabrys
Jane A. Langdale

Acknowledgements

First and foremost we are grateful to the numerous postdocs who have attended our workshops. Their enthusiastic participation has helped us refine our courses and their encouragement ensured that this book was both started and finished.

We are indebted to Debbie Alexander, Jocelyn Bell-Burnell, Alina Beltechi, Scott Crawford, Angela Hay, Julia Higgins, Steven Hill, Sue Ion, Jessica James, Peter Kalmus, Krisztian Kohary, Tim Softley, Adrian Sutton, Mark Telling and Jamie Warner for taking the time and interest to write about their own experiences and to share their wisdom. We are also grateful to Jos Schouten for the Excel calendars in Chapter 1 and for doing the index, to Wojciech Zając for the representation of the Belbin test results, to Brian Stewart for input to Chapter 22, and to Chris Trevitt for advice on Chapters 16 –18.

Matt Hodges, Jim Fouracre and Heather Sanders all read and commented on drafts of various chapters and Jill Harrison read the first complete draft – thank you all for your input!

Graham Hart and Simon Capelin at Cambridge University Press patiently extended two missed deadlines as we battled to find time to write – the lack of pressure was much appreciated. And finally our thanks to Laura Clark, Mary Sanders and Abigail Jones for competently and efficiently steering us through the editorial and production processes.

Part I

Becoming an independent researcher

You have to learn the rules of the game. And then you have to play better than anyone else.

Albert Einstein

In the first part of this book we look at the skills needed to make the transition from being a postdoc working with a principal investigator (PI) to becoming an independent research scientist. Successful careers in the sciences are normally built on good time management skills and so in the first chapter we explore how to manage your time and look at the different roles you have to play in academia. We then turn our attention to the task of communicating your research both verbally (Chapter 2) and in writing (Chapter 3). In both cases you need to be able to handle any subsequent criticism and we address how to do this in Chapter 4. We then move on to grant writing in Chapter 5. Convincing others that you and your science are worth funding is an essential skill to learn. Similarly, you must be able to manage the project to successful completion once it is funded, and so we look at tools for managing research projects in Chapter 6. Finally we close this section with three chapters that look at the 'next step'. Chapter 7 looks at alternative careers in science, Chapter 8 analyses the art of applying for an academic job and preparing for the interview, and Chapter 9 extends this analysis to applications for independent research fellowships.

1

Managing your time

The focus of your life as a postdoc is research but as a university academic you will increasingly have many more duties and responsibilities. Research becomes just one part of your life – teaching, administration, family life and interests outside of work all have to be fitted into the same amount of time. Using a 'principle-based' framework established by Stephen Covey, adapted here for life as a scientist in a university, this chapter aims to help you balance conflicting demands on your time.

The theory

If you are reading these words now, the chances are that you have become aware of the need to manage your time or have decided that your current system does not quite deliver. Time management theory has evolved over the years through first, second, third and now fourth generations – each generation seemingly improving on those that came before. In a nutshell, the first generation approach aims to bring order into chaos through organisation, and is epitomised by the use of 'to do' tools. In contrast, the second generation centres on the protection of personal time in order to be effective and is about planning and preparation, whereas the third generation focuses on prioritising goals. Perhaps not surprisingly, the fourth generation encompasses all of this and aims to harmonise personal and professional aims. Objectives are set through deep questioning and reflection, and are realised through the pragmatic use of time management tools such as weekly scheduling. Among the sources of fourth generation advice on how to cope with the increasing demands of work and the hectic pace of life are the books authored and co-authored by Stephen Covey. Here, we refer to and make use of his 'principle-based' framework (Covey *et al.*, 1994). According to his philosophy, short- and long-term goals can be achieved by developing independence and interdependence, respectively. Crucially, both are based on sound principles – life-long personal development and successful delegation where appropriate.

Habits

Covey lays the foundations for his theories in his book entitled *The 7 Habits of Highly Effective People* (Covey, 2004). Here, he describes the attributes and actions of effective, successful people. His definition of habits is an unusual one: a habit is an intersection of *knowledge*, *skill* and *desire*. In this framework, knowledge is *what* and *why* to do; skill is *how* to do; and desire is the motivation, or the *want* to do (Covey, 2004). Importantly, while you can teach skills and acquire knowledge, you cannot teach desire – you either have it or have to discover that you have it.

Covey's seven habits are grouped into two categories: those fostering independence and those fostering interdependence. Independence can be achieved by mastering three habits:

1. Be proactive – i.e. recognise actions that are within your circle of influence and act on them.
2. Begin with the end in mind – i.e. know your short- and long-term aims.
3. Put first things first – i.e. know how to prioritise objectives so that you achieve your aims.

Interdependence relies on being able to see your aims in a broad context and on recognising that more can be achieved in synergy with others. Four habits promote interdependence:

4. Think win–win – i.e. think beyond your own gains and engage in mutually beneficial relationships.
5. Seek first to understand and then to be understood – i.e. put yourself into another person's position before making a judgement, analyse arguments objectively and empathise.
6. Synergise – this will follow from mastering 4 and 5 as both will inevitably change your understanding of situations.
7. 'Sharpening the saw'.

Rather obtusely, the seventh term refers to the renewal of physical, mental, spiritual and social/emotional strength – i.e. managing your mind. Maintaining this habit is vital for long-term effective functioning and for finding a balance between your different roles; we will return to this theme in the final part of the book.

Roles

Different people have different areas of responsibility or contribution (*roles*). To excel in each of your roles, Covey advocates that each individual should be guided by a *compass* (ethical principles) and a *clock* (scheduling). The compass helps to identify the things that are most important in your life, and thus aids personal development (Covey, 2004). The clock enables you to achieve your goals. The mutual dependence of a compass and a clock is explored below.

The practice

The common (mis) perception of life as an academic is that of a dignified pursuit of truth, blue-sky research and learned discourses between fellow inhabitants of ivory towers. This was perhaps true in the nineteenth century, but modern life as an academic is as complex, busy and stressful as that of a company executive. Time is a precious commodity and there are conflicting demands that have to be resolved in order to lead a balanced life. So what can you expect, and what is expected of you as an academic in the sciences? A recent advertisement for a university position spells it out:

Preference will be given to applicants with experience and research vision in one of the highlighted areas. The appointee will be required to engage in research which will contribute to the Department's reputation; to teach, supervise and examine undergraduate and graduate students, and to contribute to administration.

There is no doubt that balancing these different demands on your time is not easy. While time management in academia is essential, it differs from that needed for business or administration. Partially, this is due to the greater flexibility that academics have in terms of hours and place of work. However, there are similarities that can be exploited if the appropriate framework is used. We have found that Covey's approach that was originally proposed for business (Covey *et al.*, 1994) adapts well to academia. The following sections illustrate how you can use this framework.

Focus on the most important things

Somewhat simplistically, the 'what' of time management requires that you identify priorities, organise time around them and keep on executing tasks as you go along. It is the 'how' that is more challenging, as even the best plans may need adjusting. At the risk of being somewhat prescriptive, the following suggests where to start (adapted from Covey *et al.*, 1994).

1. Remind yourself frequently 'why bother?' (*identify your aims*).
2. Ask yourself what your duties and responsibilities are (*identify your roles*).
3. Select quality short-, medium- and long-term goals in each role (*identify your objectives*).

4. Classify your objectives in relation to urgency and importance (*identify your priorities*).
5. Plan long term on the basis of an academic year calendar and short-term on a weekly basis (*devise your plan*).
6. Allow for the unexpected (*build in your contingency*).
7. Evaluate your week, your month, your year (*measure your success*).

In a nutshell, only do things which are either important or help you to achieve your goals. Below we look at the seven components in turn.

Your aims

The novelty of Covey's time management methodology was a proposition that the most successful people are guided both by a compass and a clock. What is a compass? In an analogy with walking, rather than wander aimlessly along roads leading from A to B you use a compass to point you in the right direction and you consult it frequently to check whether you are still going in the right direction. In this way, if you have to deal with obstacles in your direct path, you can get back on track quickly. The same reasoning applies to your personal compass – it is informed by your personal objectives (mission) – and should always point to 'true north'.

Some people get uneasy when they hear the word 'mission' because its use has sometimes been abused. However, most of us have a mission in life, though we rarely articulate it. As scientists we are more likely to think of it as our aim in life. Generally, our aims evolve and are refined as our career progresses. Importantly, harmony is only possible if your aims are aligned with the mission of your institution – check what your institution's mission is. Many universities' overarching mission statements address institutional aims first and then make provisions about its members. The same is true for professional bodies.

In the example in Box 1.1, the personal mission statement is an overarching one. It is difficult to say what this person really would do in order to 'share my teaching experience'. However, things become clearer if we are told that 'sharing experience' will be achieved by holding seminars on different aspects of lecturing, and by providing written materials for newly appointed staff, some of which may be published. In that context, the personal mission is clearly concordant with the University aims to 'achieve and sustain excellence in every area of its teaching and research . . . and to publish educational materials'. The personal mission statement then also becomes measurable.

Pause for thought: The significance of a personal mission is that it informs and guides your thoughts and actions. Write a statement of your personal aims now.

Box 1.1
An example of concordant mission statements

Personal mission statement:

To share my teaching experience and to help younger scientists achieve their potential as teachers.

University mission statement (Oxford 2008):

The mission of the University of Oxford is to achieve and sustain excellence in every area of its teaching and research, maintaining and developing its historical position as a world-class university, and enriching the international, national and regional communities through the fruits of its research, the skills of its alumni, and the publishing of academic and educational materials.

The value of knowing what your aims are is stressed in virtually every book on time management (e.g. Godefroy and Clark 1990) but it is most movingly illustrated by the case of Viktor Frankl. During the Second World War, Frankl was incarcerated in a Nazi camp but he survived due to a deep personal conviction that he must live to give evidence to the world after the war (Frankl, 2004).

While setting your aims, it is helpful to have a broader view of what it means to be a scientist (Rothwell, 2002). The recently established Researchers Portal (http://www.vitae.ac.uk/) is an important up-to-date source of information for researchers and their employers. If you are thinking about working in academia in the US, then *Tomorrow's Professor* (Reis, 1997) is a must-read. If you have already decided to follow an academic career in the UK, then Blaxter *et al.* (1998) sets the scene by describing sub-sectors of higher education institutions, and expectations of both the employer and employees.

Your roles

The most frequent roles encountered in academia are:

- Researcher
- Colleague
- Supervisor
- Author
- Teacher (lecturer)
- Friend
- Family member

Manager, mentor, personal tutor, administrator and admissions tutor are some of the other roles.

Identifying your roles allows you to manage potential conflicts of interest, spell out expectations in each role, and to organise a weekly schedule.

> **Pause for thought:** Write down your current roles now. If there are more than seven, identify seven main ones with 'sub-roles'.

Your objectives

For the purpose of planning it is important to identify one or two leading roles for any given timeframe and to identify objectives that can be achieved in that time-frame (see example in Box 1.2). To some extent any role is determined by the rhythm of an academic life. In term time, it is likely that lecturer/tutor roles will dominate a working week, whereas at other times researcher and author will be predominant.

Box 1.2		
Examples of defining objectives in the context of roles		
Role	Objective for week	Objective for year
Researcher	Generate new transformants	Get funding for rice project
Author	Start draft of Chapter 2	Co-author book on complexity
Lecturer	Complete handouts for second class	Deliver thermodynamics course

> **Pause for thought:** Choose one or two 'leading roles' for the next week, and write down the objectives to be achieved in each now.

Your priorities

Having identified objectives in a broad timeframe, you are still some way from establishing precisely when tasks should be carried out. To bring some order into chaos, it is important to classify your tasks and actions in terms of urgency and importance. The rationale behind this particular classification method is to determine, to *whom* a given activity is urgent or important. A 'time management matrix' as shown in Fig. 1.1 will then allow you to set priorities.

The list below gives an indication of the type of tasks and activities that belong to each quadrant in the time management matrix:

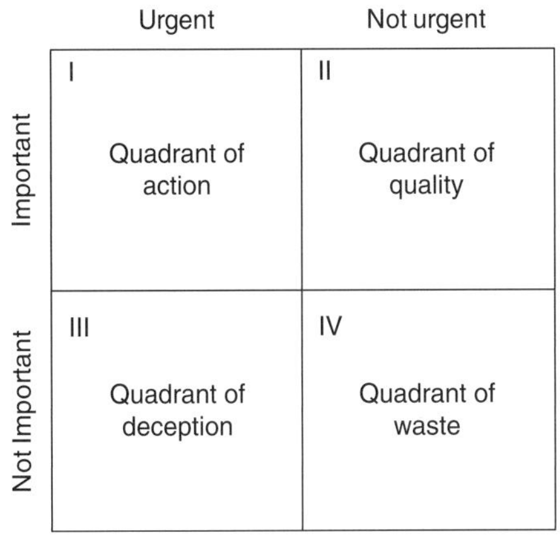

Fig. 1.1. Time management matrix adapted from Covey *et al.* (1994).

QI   – certain meetings
               – ongoing experiment
               – delivering lectures, classes, seminars or tutorials
               – submission deadlines for grant proposals
QII   – preparing a conference presentation
               – developing a new course
               – planning a new experiment
               – supervising graduate or project students
               – writing articles, reviews, books
               – spending time with family
QIII   – certain meetings
               – some emails and phone calls
               – some administration
               – commuting to work (if avoidable)
QIV   – junk email or emails that can be deleted unanswered
               – trivial pursuits, e.g. playing solitaire online or searching for spurious information
               – tweaking perfectly working experimental equipment
               – too many excuses for cigarette or coffee breaks etc.

In general, Quadrant I (QI) activities should have been scheduled well in advance, and often the times of such activities are beyond your control. A safe assumption is that, at any given time, you will spend 25%–35% of your time in this quadrant (Covey, 2004). However, the skill is to be well prepared for delivery and participation in such activities, and to avoid the urgency addiction. Although some people thrive on being constantly busy and on crisis management – life is more peaceful if you don't. Any time spent in Quadrant II (QII) on planning and preparation will more than pay off when the time comes to deliver in QI. Quadrant IV (QIV) activities are self-explanatory, and most scientists do not spend much time there. In some cases such activities can have a positive effect, allowing a tired brain to recover. However, if you find yourself spending too much time on them, it can be an indicator of stress or overwork – a sort of alarm bell. The trickiest of all is Quadrant III (QIII) – a legitimate question being: 'if something is not important, then how can it be urgent?' The key here is that QIII activities may be urgent for others but not necessarily important for you. A classical example is being asked to serve on a committee that would be worthwhile but is not a priority for you at *this* time.

> **Pause for thought:** Think about all of the tasks and activities that you are currently involved with or are planning. Classify your tasks and actions according to a four quadrant time management matrix. Remember that the level of importance should be guided by your aims (your compass).

A completed time management matrix holds the key to determining your priorities. In general QII are the highest priority and QIV the lowest but QI are the highest at their scheduled times.

Your plan

When you have a clear overview of your roles, goals and priorities you are ready to start time-tabling. There are three main ways to deal with tasks: you can *schedule*, *delegate* or *barter*. A successful delegation is possible if there is a well-qualified person or persons you can delegate to. By 'bartering', we mean exchanging tasks with somebody else – a classic example would be delivering the same lecture course to different departments while having one of yours delivered by somebody else. The same is true for teaching small groups and tutoring individuals. These two methods can free some time for other tasks.

Having worked out what needs to be scheduled, it is time to map things on to a specific time period. Any time management scheme attempts to make the best use of a clock. It is the length of time slots that vary between different methods. A very popular 'to do' list of tasks to be performed, usually drawn on a daily basis, can be counter-productive. Quite often, tasks that are not done on day one get pushed into the next day and so by the end of the week the list can still be very long. In our experience, weekly planning works well – provided you inspect the big picture first. That is, use a yearly academic calendar and first put in all fixed dates, both work and personal. Even if you do not teach or tutor at the moment, you probably feel the

ACADEMIC CALENDAR 2010-11

October '10	November '10	December '10	January '11	February '11	March '11	April '11	May '11
	Mon 1						
	Tue 2			Tue 1	Tue 1		
	Wed 3	Wed 1		Wed 2	Wed 2		
	Thu 4	Thu 2		Thu 3	Thu 3		
Fri 1	Fri 5	Fri 3		Fri 4	Fri 4	Fri 1	
Sat 2	Sat 6	Sat 4	Sat 1	Sat 5	Sat 5	Sat 2	
Sun 3	Sun 7	Sun 5	Sun 2	Sun 6	Sun 6	Sun 3	Sun 1
Mon 4	Mon 8	Mon 6	Mon 3	Mon 7	Mon 7	Mon 4	Mon 2
Tue 5	Tue 9	Tue 7	Tue 4	Tue 8	Tue 8	Tue 5	Tue 3
Wed 6	Wed 10	Wed 8	Wed 5	Wed 9	Wed 9	Wed 6	Wed 4
Thu 7	Thu 11	Thu 9	Thu 6	Thu 10	Thu 10	Thu 7	Thu 5
Fri 8	Fri 12	Fri 10	Fri 7	Fri 11	Fri 11	Fri 8	Fri 6
Sat 9	Sat 13	Sat 11	Sat 8	Sat 12	Sat 12	Sat 9	Sat 7
Sun 10	Sun 14	Sun 12	Sun 9	Sun 13	Sun 13	Sun 10	Sun 8
Mon 11	Mon 15	Mon 13	Mon 10	Mon 14	Mon 14	Mon 11	Mon 9
Tue 12	Tue 16	Tue 14	Tue 11	Tue 15	Tue 15	Tue 12	Tue 10
Wed 13	Wed 17	Wed 15	Wed 12	Wed 16	Wed 16	Wed 13	Wed 11

Fig. 1.2. A snapshot of part of an academic calendar.

pulse of university life – it is usually more difficult to get hold of people during teaching periods than outside of them. The snapshot in Fig. 1.2 shows an academic calendar for the University of Oxford where the three terms have been marked by a dashed grey stripe at the side. In your own version they can be highlighted in different colours. University closure times, weekends and public holidays are marked in grey – again, you can elaborate on this in a colour version. Our calendar shows 15 months, because academic staff often need to plan far in advance for conferences, new teaching courses, admissions etc.

Pause for thought: Fill in your academic calendar with all the known meeting/lecture/ tutorial and conference dates now.

In the framework of the big picture, a weekly plan can now be made. Looking to a week ahead, we first consider our roles and tasks (see example in Table 1.1).

We then need to ascribe the tasks and actions to their respective quadrant. This is not easy but is essential. In the example shown in Table 1.2, the main events are first listed in relation to who is organising them. This makes the urgency aspect clearer, and can allow distinction between QI or QIII. It is clear that all seminars or events belong to QI, the quadrant of action because they need to be delivered at the advertised times. Attendance at the opening of the JSPS office will provide an opportunity to find out new information about available funding

Table 1.1. *Objectives in the context of roles*

Roles	Objectives for week	Objective for year
Supervisor	Lead/conduct several seminars	Develop/deliver a series of workshops in academic practice in sciences
Researcher (sub-roles: co-author, collaborator)	Paper: write 500 words Cultivate relationships and meet new people from relevant funding agencies	Develop a new way to describe scattering in an intermediate angle range Apply for funding for UK – Japan collaboration
Family member	Go for a long walk with sister	Improve relationship with sister to sort out inheritance problems

Table 1.2. *Assignment of tasks to quadrants*

Who	What	Quadrant
Higher Education Academy (HEA)	Changing practice in engineering	QIII
Japanese Society for Promotion of Science (JSPS)	Opening of new London office	QII
MPLS Division meeting	Training registers	QII/QI
MPLS Division event	Becoming an independent researcher	QI
Oxford Learning Institute event	Supervising PhD students	QI
Oxford Learning Institute event	Examining PhD Students	QI
MPLS Division (Leading seminar)	Coping with complexity	QI
Myself – co-author	Writing paper on short range order	QII
Myself	Clear office	QIII
Myself – collaborator	New data reduction	QII

and will both renew old contacts and establish new ones – a typical QII activity of relationship building. While the conference organised by HEA is interesting, it is not essential to attend for our planner, hence QIII assignation. The meeting on training registers is important and urgent (QI) due to a need to set long-term divisional objectives (QII).

Of the QII events, it is useful to identify the most important role and objective for the week and then schedule time for that. While we cannot command creative thinking at will, we can create opportunities for it in our schedule. Other QII events, whilst important are by definition not urgent, and it is these events that can be shifted if necessary to allow for the unexpected.

Week of 9th March 2008

> **The most important role and goal this week is:**
> author – write paper on SANS and WANS

		Sunday	Monday	Tuesday	Wednesday	Thursday	Friday	Saturday
am								
	9					QII	QII	QII
	10		QIII		QI–OLI	writing	preparation	problem list
	11		HEA conf.		QI			
	12		HEA conf.	QI	QI			
	1	*lunch*	*lunch*	*lunch*	*lunch*	*lunch*	*lunch*	*lunch*
pm	2	QII writing	travel	QII writing	QIII – data	QI – OLI	QI	
	3	QII					managing	
	4	QII	QII – JSPS				your time	
	5	QII		QII writing	QIII – clearout		QI	
	6	QII			QIII – clearout		week's evaluation	
eve	7							
	8							
	9							

> **Manage your mind – to start or continue this week (what and when):**
> start Tai Chi 15' daily from Tuesday

Modelled on S.Covey's weekly worksheet (*The Seven Habits of Highly Effective People*)

Fig. 1.3. A completed weekly spreadsheet.

Having assigned tasks to quadrants, a weekly plan can be generated by dividing days of the week into four time chunks with hourly timeslots: morning, lunch, afternoon and evening. A weekly spreadsheet can then be filled in as shown in Fig. 1.3.

A comment on the process of filling in the weekly spreadsheet is in order here. Covey offers a nice analogy about the necessary sequence. Imagine you have a jar to be filled with large stones, pebbles, sand and water. Clearly the sensible thing to do is to start with the largest items (large stones), fill in the remaining spaces with progressively smaller ones and finally add water. If we reverse this procedure, we risk not being able to fit in large stones at all (Covey *et al.*, 1994).

> **Pause for thought:** Think of either a typical or specific week and fill in a weekly spreadsheet now. A copy of a clear spreadsheet can be downloaded and updated from our website. Other electronic calendars are also available – Google Calendar (http://www.google.com/calendar) and Microsoft Outlook both allow others to access your schedule (with either editing and/or viewing rights).

Your contingency

Everybody knows the sinking feeling when the deadline for the submission of a grant proposal approaches fast, and concurrently an opportunity arises that if not dealt with immediately will be missed. How then do you decide priorities and make space in your schedule? The key here is to recognise that effective planning is just a

Week of 9th March 2008

The most important roles and objectives this week are:
author – write paper on short range order
family member – visit sister

		Sunday	Monday	Tuesday	Wednesday	Thursday	Friday	Saturday
am	9					QII	QII	QII
	10		QIII		QI – OLI	writing	preparation	problem list
	11		HEA conf.		QI			
	12		HEA conf.	QI	QI			
	1		*lunch*	*lunch*	*lunch*	*lunch*	*lunch*	*lunch*
pm	2	family visit QII	travel	QII writing	QIII – data	QI – OLI	QI	QII writing
	3	family visit QII					managing	QII writing
	4	family visit QII	QII – JSPS				your time	QII writing
	5	family visit QII		QII writing	QIII – clearout		QI	QII writing
	6	family visit QII			QIII – clearout		week's evaluation	QII writing
eve	7							
	8							
	9							

Managing your mind – to start or continue this week (what and when):
start Tai Chi 15' daily from Tuesday

Modelled on S.Covey's weekly worksheet (*The Seven Habits of Highly Effective People*)

Fig. 1.4. An example of shifting activities within the week's frame.

road map and is not set in stone. We should welcome challenges and pick up opportunities when we can. Importantly, the more time we have scheduled for QII activities, the more space we have (both mentally and physically) to allow for contingency. In the example shown in Figure 1.3, our researcher got so focused on writing a paper that the QII objective of spending more time with family was forgotten and is missing. A phone call from a distressed sister resulted in a change of plans, as shown in Fig. 1.4 (compare entries for Saturday and Sunday in both figures). Going for a long walk with her sister also brought in an element of physical activity (sharpening the saw).

Pause for thought: Keep monitoring your weekly schedule during the chosen week and adjust as necessary.

Your success – evaluation

An essential part of the planning process is to evaluate each week before planning the next one. The main point is to revisit all of the set goals, decide whether they were fulfilled or not, think about progress and then feed the conclusions into the planning for next week. If things did not go according to plan, think about what went wrong so that you can learn from your mistakes. If things went well or exceeded your expectations, give yourself a pat on the back and a small reward.

> **Pause for thought:** Evaluate your week and use the conclusions drawn to plan another week ahead.

Why a weekly and not a monthly evaluation? The following argument may clarify it. Prioritisation of aims needs to be done in the right timeframe. It can be likened to the invention of perspective in painting. Byzantine icons are 'flat', there is seemingly no order to them, objects have strange proportions – e.g. people larger than horses, and all objects are in the same plane. Introduction of perspective brings order: objects take on their correct dimension and place. When we take a step back from the picture, we see it in its totality (weekly plan). When we go further away from the picture, we see it in the context of the surroundings (yearly plan). When we take a step forward, we can see minute detail (one day). Hence a week provides balance between long-term planning and the minutiae of daily work.

As a test example for the above methodology, we present a case study in Box 1.3. We encourage you to think about it and discuss possible outcomes with your colleagues.

Box 1.3
Case study

Emily and Peter met at the International Conference on Climate Change in 2007. They both work and live in the UK – Emily is on her second postdoc at the University of Oxford and Peter is on a prestigious 5-year fellowship at the University of Edinburgh. They are renting a house in Edinburgh, with Emily staying in a rented room four days a week in Oxford. She normally manages to catch up on reading on train journeys, and while in Edinburgh she can answer emails.

This particular week things are not going smoothly for her. She needs to submit a paper by the end of the week in order to meet a conference deadline. While she was looking forward to a long weekend in Scotland, her supervisor asked her to look after a couple of graduate students in his absence. One of them has to start a difficult experiment and clearly has no clue how to begin. The second one is panicking since there is a deadline approaching for the submission of his thesis. She has also been asked to referee a grant application (with very little advanced warning) and feels it is a first sign of being recognised as a scientist in her own right. If she delays or postpones her trip to Edinburgh, Peter is likely to be very upset since they planned a 3-day trek a long time ago.

What is Emily to do?

Five tips for more effective time management

As you may suspect by now, there is neither a magic formula nor 'one size fits all' collection of recipes for managing your time. However, we find that the following *MODEL* can help:

- *Maximise quality time spent on QII activities*. Make an appointment with yourself and treat this time as you would honour an appointment with another person. Switch off phones, turn off email software and remove as many distractions as possible. Stay at home if there are fewer distractions there.
- *Only tackle jobs once*. When a piece of paper or email lands on your desk, skim read it and decide whether you are going to deal with it immediately. If you are not, decide how long it is likely to take (or better still how long you are prepared to spend on it) and then schedule a time to do it (this could be later in the same day or in a months time). Do not revisit (mentally or physically) until the scheduled time. This approach prevents you fretting about the fact that it is outstanding. Once completed, file/destroy/delete documents as appropriate.
- *Delete as many emails from your inbox as possible*. Use subject headings as an indicator of whether a message is relevant to you. If it is not relevant delete it immediately. Once relevant messages have been responded to then delete them or move to a dedicated folder.
- *Ensure your working space is tidy*. It is difficult to estimate the amount of time wasted in search of vital documents/reagents/equipment.
- *Let your biological rhythm influence how you plan your activities and tasks*. Remember that lack of sleep impairs performance. Schedule QII activities for when you are most creative and QIV for when you are least efficient.

> **Pause for thought:** Think of your own time-saving devices and write them down now.

How we did it

At the time of writing, this workshop is offered university-wide twice a year. It lasts about two and a half hours because the participants are asked to complete the exercises that are scripted here as 'pause for thought'. They are expected to try adapt the methodology to their specific circumstances and are encouraged to have frequent small group discussions about the validity and ease of this approach. To date, we have had very positive feedback from several participants who have incorporated this approach into their daily life.

Summary

Successful time management depends on the recognition of what we want to achieve and of how we want to achieve it. The route to success is guided by a compass of principles, and punctuated by a clock of weekly schedules. To increase the effectiveness of our actions a systematic approach is needed. This can be achieved by first assigning tasks a quality and action value, and by maximising the time spent on important tasks. Evaluation of plans on a weekly basis reinforces this approach.

Selected reading

If you only have time to read one book, make it:
Covey, S. R., Merrill, A. R. & Merrill, R. R. (1994). *First Things First*. London, Simon & Schuster.

Other texts:
Blaxter, L., Hughes, C. & Tight, M. (1998). *The Academic Career Handbook*. Buckingham, Open University Press.
Covey, S. R. (2004). *The 7 Habits of Highly Effective People*. London, Simon & Schuster.
Frankl, V. E. (2004). *Man's Search for Meaning: The Classic Tribute to Hope from the Holocaust*. London, Rider.
Godefroy, C. & Clark, J. (1990). *The Complete Time Management System*. London, Piatkus.
Reis, R. M. (1997). *Tomorrow's Professor*. New York, IEEE Press Wiley-Interscience.
Rothwell, N. (2002). *Who Wants to be a Scientist? Choosing Science as a Career*. Cambridge, Cambridge University Press.

Web resources
Howard Hughes Medical Institute – Making the Right Moves: A Practical Guide to Scientific Management for Postdocs and New Faculty. http://www.hhmi.org/resources/labmanagement/

2

Giving a good research talk

There is only one steadfast rule to ensure that your research talk is a success: 'know your audience'. Occasionally, you will give research talks to very specialised audiences who are desperately interested in your science and will listen even if you speak to your feet. However, most of the time this will not be the case and you should be aiming to entertain as many people as possible. There is no better advert for your research programme than someone outside of your field telling someone else that they heard a great talk by X on Y. This chapter provides guidance on how to engage your audience.

The theory

Explaining your research to your peers, funding councils, other stakeholders and the general public is a professional necessity. Unfortunately, however, scientists do not enjoy the best reputation as communicators even when giving a research talk aimed at their peers. There are plenty of books and websites giving advice on how to use packages such as PowerPoint, how to decide about the content of a presentation or how to keep the audience interested using multimedia (Meredith, 2010). These sources will tell you how to improve your performance, to keep to time and to deal with the questions during the talk or after it. This is all very useful and important but we think that the key to a successful presentation is your engagement with your subject and your audience.

If you go to a live performance by famous musicians, you might notice something – invariably the music speaks through them, their bodies and instruments. The best ones are no longer there in some sense; they are lost, immersed; they channel the music. They are not preoccupied with how they look or what the audience thinks about their performance – the only thing that counts is living the music – becoming the music. Much the same can be said about a research talk, no matter what the subject is. The best speakers we have heard were 'hardly there' – all that counted was their subject; they were the subject.

So how do you engage an audience? An essential prerequisite is a clear, well laid-out presentation. Then comes the performance. Most people will listen if you look as though you are enjoying yourself. What are the general characteristics exhibited by someone having fun? – they smile, they become animated, they often raise their voice, they vary their tone and they make eye contact. Think how animated you become when you have a great conversation with an interesting person. In contrast, someone who is uncomfortable will mumble, close their body, fidget and look away. In seeking to engage therefore, one of the most important skills to master is voice projection, since it is almost impossible to be monotonic when projecting. And practice, practice, practice your equivalent of music scales so that your performance is so smooth that is becomes invisible. Becoming one with your subject will then be automatic.

The practice

Before thinking about how to give a good research talk, perhaps you should take one step back and ask yourself a question: why bother? In more polite language this means: why would I want to give a talk? Is this the right time for me to do so? Will my ground-breaking research withstand the onslaught of competition or is it too early? Who should I talk to – my departmental colleagues, my university, other stakeholders, my country or the world? These questions may seem far-fetched – but in the age of the internet there is no hiding. As soon as the subject of your seminar is posted on the group or departmental website, it becomes visible – it can be 'Googled' (unless the website is password protected). Most likely, the seminar organiser will send an email around with the abstract of your talk trying to whip up some support and get busy people to come and listen to you. It is like a stone thrown into water – the ripples spread far away. So, the main reason for giving a talk is to get you and your science recognised. Of course, it will only be recognised positively if you give a good seminar – so how can you maximise the chances of that happening?

Type of talk

First you should think about the type of talk you would like (or have) to give.

Working group seminar

Progress talks to group members can often be harder to give than research seminars. This is because it is easier to talk about the completed part of a project because you can look back and recognise key stages. This type of perspective allows you to blur the details and to give a broad overview. In contrast, with progress talks you may not yet know what the crucial stages of the project are and you may subconsciously worry about surprises in store for you. Given the informal nature of group talks, you may not want to spend time making a PowerPoint type presentation. However, you still need to be clear about the relative importance of the points you make and your talk should still have a clear beginning, middle and end. So make good notes and use a white- or blackboard alongside a relevant selection of photos, video snippets or animations. You can expect this type of talk to be interactive, with your colleagues asking questions and chipping in as you go along.

> **Pause for thought:** Think about a project you are currently working on. Sketch out the main points and think how you would illustrate them to the best effect.

Research seminar

You may be invited to give a talk to your department or at another university. Depending on the circumstances you may be asked to talk to a specialist or to a generalist audience. If you are speaking to specialists, they will be interested in technical details whereas a general audience will appreciate it if you present your work in a wider framework. In the US, it is common to organise a tour giving talks at several universities, and it is possible to do the same thing in Europe. These tours serve two purposes: they let others know about your research and they allow you to learn about research being carried out elsewhere. With each successive talk, you will see a remarkable improvement in the quality of your presentation. Such tours also give you a chance to fish for academic positions, and when you make a good impression you are likely to be invited to apply when a position does becomes available. Hence you should treat every research talk as an interview audition.

Conference seminar

This type of talk will have a slightly different focus depending on whether you have a contributed presentation or are invited to give a plenary talk. The research you present will also be determined by the subject of the conference. You should always have a look at the programme to see where and how your talk would fit in. You do not want the embarrassment of having to skip half of your slides because the previous speaker has just told a large part of your story. Generally, at conferences you can expect the audience to comprise both experts and people who want to get into the particular research field. As such, you will not need to spend much time on background information, instead you can focus on your contribution to the field. Even with good preparation, there are a number of factors that can adversely influence the delivery of your talk at a conference: the most important being time-keeping – or the lack of it – by others. Having a good chair is not a given, so if the previous speaker has droned on, you may find your time squeezed. Annoying as this is, stick to your well-timed talk as much as possible and be prepared to cut some of it if necessary.

Public seminar

This is possibly the most difficult type of talk to give, especially if your audience is a group of curious 10-year-olds. Such talks used to be given only by distinguished scientists but institutions and individuals everywhere now recognise that it is important for scientists at all career stages to inform and excite the public about scientific research. As always, you must know your audience – there is a big difference between talking to school children and talking to retired members of the Rotary Club. In both cases, however, the technical details should fade into the background and your slides should be attractive, with plenty of pictures, only a few diagrams and hardly any formulae. After the talk, be prepared for unusual questions and be ready to think on your feet – always repeat the question as it gives the audience a chance to hear it, and gives you a moment or two to think about the answer.

Pause for thought: How would you present your current research to a group of 14-year-olds?

Preparation

Once you know what type of talk you have to give, you can start the preparation. And remember – the more polished the product, the more work has been done behind the scenes. Think about it in terms of building a skeleton and then dressing it up. You do not want to show the skeleton in your talk, but if it is not properly assembled then the clothes will not fit.

When you start working on a presentation, it is important to keep the purpose of it in mind at all times. Are you going to talk on an emerging subject, hot off the press with many questions to answer, or are you going to describe part of a research project that has a start, middle and end? Even when you have decided on the purpose, it is not easy to stay on track because there are so many interesting things to say and you may feel that you need to support everything with evidence (i.e. details).

Regardless of the type of talk and precise purpose, its success will depend mainly on three factors:

- good science
- good storyline
- good delivery

Given these three absolutes, there are eight other points to consider when you start preparing your talk. The first four will have been dictated by others, whereas you have some control over the last four.

Pause for thought: Think about the research talks that you have attended recently. Which ones were memorable and why?

The audience

Who are you giving your talk to – your research group, scientists from your department, a group of international scientists at a conference, the general public? Whilst there are clearly different levels of engagement between these groups, the edges are fuzzy – your colleagues will benefit from a clear explanation of a complex phenomenon stated in simple terms, normally 'reserved' for the general public. Similarly, there is no doubt that school children and the general public will benefit if we can explain complex scientific data in an exciting and comprehensible manner.

The content

You may be asked to speak on a given subject, or have a free choice of topic. In the first case, stick to it, in the second case, choose a topic that you are really enthusiastic about.

The venue

Is it a small informal seminar or a major international conference? This should inform your choice of presentation media. Software packages such as Open Office, PowerPoint or Keynote (for Apple Mac) are expected in most cases, but for informal seminars you could use a whiteboard and marker.

Your character

Are you a born show person or do you dread public speaking no matter how small the audience? Obviously, we cannot change your character but there are some things which can help – the first and most obvious being practice. On a more basic level, learn to relax your body – when we are nervous our muscles tighten and as a consequence our throat closes – not good for voice projection. Taking a few deep breaths before speaking helps you to loosen up.

Key points to make

No more than two or three per talk.

Organisation of the talk

Title, outline, clear indication of each topic and subtopic, conclusions, acknowledgments.

Timing

This is absolutely crucial. How often have you fidgeted in your seat at a conference because somebody went overtime? If you are giving a job talk, the ability to keep to time will score well with the interviewers and will show your professionalism and respect for other people's time.

Slide details

Trim all fancy features and concentrate on simple, well-laid-out slides. The background should be as neutral as possible, white is fine. Think about your audience in the back row of a non-theatre-style room – will they be able to see the bottom part of your slides? Will they be able to see anything at all if the font is 10 point? How big should the figures be? What colours should be used? – remember 8% of men and 1% of women are red–green colour blind.

> **Pause for thought:** Have a critical look at your recent research talk. Does it tell a story? Identify the key points – are they clearly made? Change all font size to 24 points – how much of the original text can you keep on the same slide? And if the font size is 30?

Examples of slides made for a group seminar and for an invited talk are illustrated in İşsever and Peach (2010).

Presentation

The voice

You have put together your talk, bearing in mind your audience, checked that there are no typos on your slides, your colours look great, the story flows – what next? You have to turn your attention to the 'performance'. Apart from the obvious points – don't mumble, don't talk to the screen, maintain eye contact with the audience – the most important one is *voice projection*. Too many scientists speak in a monotone – and thus do not convey the excitement of their subject. Non-English speakers have even more of a challenge as accent, pronunciation and modulation of voice often makes projection even more of an issue. If you are a non-English speaker, it may be worthwhile taking voice coaching classes in order to improve your projection. The bottom line is that you have to speak from your belly and not your throat to make an impact. That means you have to breathe using your diaphragm, and in order to do that, your body has to relax.

Practice

Given that you now have the slides and you have the voice – it is time to say it to yourself – in the shower, walking in the park, driving in the car – who cares where – just do it. This will allow you to check the timing and adjust the stumbling

Box 2.1									
Evaluation form for presentation skills workshop									
Speech	Pepe	Jill	Marcel	Haruko	Angela	Julia	Nicky	Paolo	Stephen

		Pepe	Jill	Marcel	Haruko	Angela	Julia	Nicky	Paolo	Stephen
Volume:	Projected									
	Normal									
	Mumbling									
Tone:	Monotonic									
	Varied									
	Fearful									
	Confident									
Speed:	Too fast									
	Too slow									
	Just right									

Face		Pepe	Jill	Marcel	Haruko	Angela	Julia	Nicky	Paolo	Stephen
	Smiling									
	Serious									
	Enthusiastic									
	Fearful									

Eye contact with	Pepe	Jill	Marcel	Haruko	Angela	Julia	Nicky	Paolo	Stephen
> 1 audience member									
1 audience member									
No eye contact									

Hands		Pepe	Jill	Marcel	Haruko	Angela	Julia	Nicky	Paolo	Stephen
	Stationary									
	Fidgeting									
	Extending									

Body		Pepe	Jill	Marcel	Haruko	Angela	Julia	Nicky	Paolo	Stephen
	Open									
	Closed									
	Static									
	Animated									

Information on slides	Pepe	Jill	Marcel	Haruko	Angela	Julia	Nicky	Paolo	Stephen
Too much									
Too little									
Too vague									
Too detailed									
Just right									
Were you entertained?									

points. You can then ask willing colleagues to listen from the back of a lecture theatre. Can they hear you and see the slides clearly? Can they remember one key point from your talk? Their honest comments may hurt but will definitely improve your presentation on the big day.

Technicalities

The technical side of a presentation can be both comforting and a nightmare. Will you be able to use your laptop (the best option) or will you have to rely on what the seminar or conference organisers provide? In both cases you may have problems with the compatibility of your platform/software/hardware with theirs – so check carefully in advance. Videos and animations are particularly prone to disaster when transferred to another computer. To limit problems, always have a back-up of your talk as a PDF file – it will work on any platform even though you might lose some of the animated features of your original talk – better that than nothing. Colours and contrast can also be altered by data projectors so go for safe options – the W3C consortium has guidelines for user-friendly web colours; these guidelines work for slide presentations as well – (http://www.w3.org/TR/UNDERSTANDING-WCAG20/visual-audio-contrast7.html).

How we did it

In our workshops, participants gave 15-minute talks about their research. All talks were video-recorded and saved on DVDs. Participants were asked to evaluate all of the talks using guidelines shown in Box 2.1, both immediately after the seminars, and again after watching the DVD. Having done that, they each came up with a selection of terms to describe what they considered to be best practice. Assessments were emailed just to the course organiser and not to the whole group. There were two reasons for this: firstly, it is easier to be frank without face-to-face 'censorship'; secondly, conclusions could not be influenced by others. The evaluations were brought together, common points as well as differences outlined and a short discussion conducted at the next workshop.

Summary

Giving research talks is a core activity for scientists. For those talks to make a positive impact they need to be delivered in a way that engages the audience. Such engagement requires careful preparation and an entertaining performance. The keys to success are clear slides, a polished delivery (gained through practice), a varied and projected voice, and eye contact.

Selected reading

If you only have time to read one book, make it:
Meredith, D. (2010). *Explaining Research: How to Reach Key Audiences to Advance Your Work*. Oxford: Oxford University Press.

Other texts:
İşsever, Ç. & Peach, K. (2010). *Presenting Science: A Practical Guide to Giving a Good Talk*. Oxford: Oxford University Press.
Lagendijk, A. (2008) *Survival Guide for Scientists*. Amsterdam: Amsterdam University Press.

3

Writing a quality research paper

A quality research paper requires more than good data. This chapter provides guidelines for best writing practice. The steps involved are illustrated both in principle and with examples from published papers. The aim is to highlight the fact that writing a good scientific paper demands more than the ability to write good english. Practical advice is provided on how to handle different types of writing projects (research article, review, book chapter or book).

The theory

The invention of a printing press by Gutenberg laid the foundation for distributing information to many readers at the same time. The next step towards the wide dissemination of scientific papers came in the second half of the seventeenth century when the Royal Society started to publish the *Philosophical Transactions* (Russell, 2010). A statement attributed to John Ziman asserts that 'the "size" of science has doubled steadily every 15 years. In a century this means a factor of 100. For every single scientific paper or for every single scientist in 1670, there were 100 in 1770, 10 000 in 1870 and 1 000 000 in 1970.' There are currently around 2000 publishers in science, technology and medicine and the estimated number of papers published in 2006 was over 1.3 million (Björk *et al.*, 2009).

Pause for thought: If there are over 1 million scientists today – what does it imply for the readership of your paper?

The evolution of professional science writing over the last two centuries resulted in the current rigid paper format known as *IMRAD* (*I*ntroduction, *M*ethods, *R*esults, and *D*iscussion) that was formalised in the second half of the twentieth century (American National Standards Institute 1979). The Introduction describes the problem studied; the Methods tell us how the problem was investigated; the Results report the findings; and the Discussion explains what the results mean. Keeping to this structure benefits both authors and referees because consistency helps to organise thoughts. Some journals, however, have a slightly different order where the Methods can be the last section or be listed in figure captions.

The practice

There are several reasons why you need to publish a research paper, the first conveyed by the saying 'publish or perish'. Making your research results widely known and available is one of the major requirements for securing promotion and for getting the next grant. Perhaps more importantly – it feels good when your paper is published – it is the completion of a project that may have started years earlier with a vague idea generated over a few drinks with colleagues. Given the professional and personal benefits of publishing it is perhaps surprising that writing often tends to be postponed to make way for other more urgent demands on your time. Writing can be a very enjoyable and fulfilling experience when given enough space in Quadrant II (see Chapter 1).

Assuming that we can create quality time for writing, lets look at how to go about it. To begin with, we need to pose four questions:

- When?
- Where?
- What?
- How?

When to publish?

In our experience the best papers are written as soon as possible after the main body of work is completed. The work is fresh in your mind and any experimental set-up or materials are still available if it becomes necessary to complement or extend experiments (either during writing or in response to reviewer's comments).

Where to publish?

The increasing number of scientists and the competition to get papers published has resulted in the proliferation of scientific journals. Original research is mainly published as papers or letters in one of these many journals and the first task before writing a paper is to decide where you want to submit it. Every discipline has got a pecking order and you certainly have an idea of which journal is top in your field. A simple quantitative measure is the *Impact Factor* (*IF*). This is the average number of citations for a journal in any given year, based on the previous two years. You can find the impact factor for most journals in Journal Citation Report (JCR), a product of Thomson ISI (Institute for Scientific Information). However, it is important to remember that even if your paper is published in the most prestigious journal it does not necessarily mean that it will contribute any more to knowledge than a paper published in a more humble journal. Citation frequencies are subject to the vagaries of fashion and often vary widely between disciplines, and yet sadly both citations and IFs are often used to measure the 'worth' of a scientist.

So how do you decide where to publish? The bottom line is to think about your preferred readers. If you want a broad audience then you should submit to a general journal. If you want to report a new technical advance that will be of major interest to a particular group of people, you need to submit to a journal that those people will read. Having decided – generalist or specialist – you then need to think realistically about the impact your work will have on the field and submit to the journal at the most appropriate place in the ranking. Of course, it never hurts to aim high if you are prepared to spend the time resubmitting elsewhere if the paper is rejected by your first choice journal.

What to publish?

Here we urge you to be selective. Both experimental and theoretical data have to be significant, clearly explained and reproducible. Less exciting results can be included, however, if they complement or contradict results or theory developed by others. Of course, in deciding what to include in the paper, you will also need to decide who the co-authors will be, and the order in which their names should appear. This is an important and often delicate task because in multi-author citations only the first name will be quoted.

How to write a research paper

In the following section we give a summary of our view on how to write a research paper. There are other writing guides available that are either condensed (Lagendijk, 2008) or written in a more conversational style (Day and Gastel, 2006).

Having decided which journal to submit to, you need to obtain a copy of the instructions to authors. This will provide information about required formats for figure/data presentation, imposed word limits and reference styles. Regardless of journal, however, a typical paper will have the following general outline:

- Title
- Authors and affiliations

- Running title
- Keywords
- Abstract
- Introduction
- Methods (sometimes after Discussion)
- Results
- Discussion
- Conclusions (not in all journals)
- Acknowledgements
- References
- Tables
- Table legends
- Figure legends
- Supplementary Information

The order in which the parts of the paper appear is not necessarily the order in which they are written. While ultimately the IMRAD order has to be respected, the paper can be written in almost any order and quite often the Introduction and Abstract are written last.

In our view, the Results should always be written first. Assembling each figure or dataset and summarising the main point it makes in a single sentence is a good way to start. If you cannot summarise the point in a sentence, maybe there isn't a clear message? Maybe there is too much data in the figure and you are trying to make too many points at once? Going through the process of assembling one dataset/one point will allow you to spot any missing data that is needed to make your point, and/ or any superfluous data that doesn't contribute to the point you are making. It will also help you arrange the data in a logical order.

Having assembled your figures/points, you are ready to decide how to convey your 'story', i.e. to decide what context to set your results in. This will ultimately depend on your prospective readership. Having decided on the context, you are then ready to write the paper. Below are a few things to consider for each section.

Title

It is often quite a challenge to write a good title. The best titles are as short as possible, yet contain all the relevant information. For example, the title 'Ecology: Life after logging' alerts us to the environmental consequences of logging forests but does not tell us where the studies were conducted (Edwards, 2010). A more appropriate title would be 'Ecology: Life after logging in Borneo'. In contrast, titles that are too long and too specific deter a potential reader. For example, only specialists are likely to look at a paper entitled 'A study of the ecological consequences of logging in the peat swamp forests of Borneo over the four year period from 2000 to 2004 shows that *Paralaxita telesia* populations are in decline'.

Abstract

The abstract should convey the question asked, the answer to the question and the significance of the work. Remember, it is likely that over 95% of the people who are drawn in by the title of a paper will only read the abstract. Box 3.1 illustrates a good example of an informative abstract.

Box 3.1

Example of a good abstract

We have added a new dimension to measurements of thermal-neutron scattering by constructing, at the Oak Ridge High-Flux Isotope Reactor, a triple-axis spectrometer with polarisation-sensitive crystals on both the first and third axes. With this instrument the distribution of scattered neutrons from an initially monochromatic, polarised beam is measured as a function of angle, energy, and spin. The usual polarised-beam instrument is a two-axis diffractometer in which the measured cross-section involves integration over the final energy and spin distributions. Information is lost in these integrations. We are able to measure cross-sections for scattering from an initial state of specified momentum and spin to a final state of specified momentum and spin. In this paper we present a calculation of the appropriate cross-sections and a series of experiments designed to explore the capabilities of this instrument and to demonstrate various applications of neutron-polarisation analysis.

(Moon *et al.*, 1969)

Pause for thought: Analyse the paragraph in Box 3.1. What are the key points?

Our analysis: This paragraph gives a roadmap through the whole paper. The first sentence states what, where and how the equipment was constructed. The second sentence sets out what is measured, and the third sentence compares this new method with the existing one. The fourth sentence explains the drawbacks of the old method. Hence 'a new dimension' is clear. The fifth sentence outlines a new type of measurement. The sixth sentence summarises the results and discussion.

Introduction

The introduction is like a shop window – it should be uncluttered, display the most essential items (i.e. introduce the characters in your story) and draw your reader in. Importantly, the introduction should introduce the issues that are raised in the discussion. It should not necessarily introduce the real reason why the work was done (which may be irrelevant to the story being told).

Methods

This section should be concise yet contain all of the information required to enable someone else to replicate the experiment, e.g. about materials, the way samples were prepared, experimental conditions such as temperature, pressure, magnetic field. It is customary to refer to an existing method with a reference to a key paper or a textbook, rather than writing out the whole protocol. However, don't be too mean with your information. For example, a few extra words can save the reader having to trawl through another paper to find out what grid size you used or whether your sections were embedded in wax or plastic. It is worth consulting papers in the journal you aim to submit to in order to assess the amount of detail expected.

Results

Having assembled figures/datasets with a point for each (as above), the results section is written by simply articulating those points, using one or at most two paragraphs for each figure. Each paragraph should make just a single point. Each paragraph should start with a sentence introducing what is coming in the paragraph and the last sentence should state that point. If this is done properly, the reader should be able to get the message of the paragraph by reading just the first and last sentence.

Discussion

The length of the discussion will depend on the nature of the results presented. If the results are essentially qualitative, there is little scope for controversy. In this case, the discussion can be relatively short. Quantitative results require a more subjective interpretation and as such, any interpretation is more likely to be questioned. In this case, the discussion is likely to be longer than the results section as an argument has to be presented and defended in the context of other published material. In both cases, the discussion should be used to suggest how the results *as a whole* have influenced the field (leaving the discussion of individual results to be presented in the results section).

Conclusions

Not all journals require (or allow) this but it should be brief and state the most important findings or applications. The example in Box 3.2 summarises what type of instrument (small angle light scattering, abbreviated to SALS) was developed and for what purpose (soft matter studies). The technical innovation of using a CMOS (complementary metal oxide semiconductor) was noted; the application to a generic

---

Box 3.2

Example of conclusions in a paper describing a new technique

A SALS instrument with high resolution at low angles and a high signal-to-noise ratio has been developed, for the purpose of a soft matter study. The use of a CMOS image sensor with a large exposure area as the detector has provided a wide range of scattering vectors without using a condenser lens. In summary, we have shown that our SALS instrument is especially suitable for the investigation of the phenomena, described by characteristic power laws, occurring in solutions and very thin films.

(Nishida *et al.*, 2008)

---

class of substances (solutions and very thin films) was highlighted. Abbreviations were defined in the paper where they first appeared.

Pause for thought: Write down conclusions for the draft paper you are writing now, even if it is incomplete. Note the most important points and check whether they are covered in the paper.

Acknowledgements

These need careful consideration! Always acknowledge a funding agency or a sponsor. Then thank your colleagues who contributed to the paper through discussions or by contributing to the work in a manner that is too minor to merit co-authorship. Good manners require that you let them know in advance that you are doing this. Sometimes people may not wish to be acknowledged if they think that their contribution is too small (forget it) or too large (they want to be co-authors).

References

All references should be chosen for strict relevancy to your paper (not too many, not too few) and written in the style of the journal you are going to submit to. The latter requirement can be a painful experience if your paper is rejected and you have to submit to another journal. The best, though rather expensive, remedy is to use a software tool such as EndNote™ for Word or BibTex if you use Latex (free). Conversion from one format to another then takes a matter of seconds.

Other formats

Letters

Thus far we have concentrated on writing a full research paper. This is likely to be the most important type of publication on your CV but in some research fields short letters take precedence in terms of prestige, e.g. *Physical Review Letters*. Such letters are reserved for first claims of ground-breaking research, and you are expected to follow a letter by a full paper in another journal. There are severe restrictions on space available for letters and they contain much less detail than a full paper. You therefore have to be ruthless about deciding what to include – there is no room for non-essential data (or words). A word of warning: because of competition for space (and prestige), letters can be refereed ferociously. Do not be disheartened by the response to your ground-breaking discovery – persist until it gets published.

Reviews

As your career progresses, it is likely that you will be invited to write a review or a chapter for a book. There are no general rules about when to accept such invitations, but you should bear in mind that these are very time-consuming activities with relatively little return in terms of credit on your CV (although they are often well cited). However, writing a review can be very worthwhile if you want to gain an overview of a new research field or to signal your area of expertise to other scientists. Perhaps you have recently done an extensive literature search in your field and feel it would be sensible to get some mileage out of that. In that case, you need to consider which journal is a suitable one and determine the type of review that is published in it. Do you want to write a critical evaluation of publications in a given field or review a compilation of papers published in a given year? Usually, original research (not published elsewhere) does not feature in such contributions but a good review can offer original insights. The US National Academy of Sciences recognises this fact and offers an annual Award for Scientific Reviewing (see example in Box 3.3).

Box 3.3
An outline of a review paper on 'Jumbo bacteriophages'

Contents
Introduction
Genome structure and evolution of medium phages
Big and bigger phages
Capsid size and genome size
Jumbo phages
Conclusions
References

(Hendrix, 2009)

'Hendrix's reviews, overviews, and minireviews have focused research in the areas of structure, assembly, and genomics of bacteriophages and include numerous original and provocative ideas.' (http://www.nasonline.org/site/PageServer?pagename=AWARDS_scirev)

There are no strict rules on how a review should be written. However, an outline that is often published as Contents at the beginning of a review helps the reader to navigate and the author to organise the writing. In comparison with a research paper, a review has to be written in more general terms as it addresses a wider scientific audience. Hence, clarity and good style are even more important than in a research paper.

Pause for thought: Look up the literature section of your PhD thesis. Would it qualify as a minireview? Why (or why not)?

Book chapters

Essentially the same guidelines apply as for writing a review – deciding detailed scope, careful planning and writing an outline are the first steps. However, the bigger the book and the more contributing authors, the more likely publication will miss the deadline. Bear this in mind and try to provide your chapter in a timely manner to minimise delays. If delays are too long, information will already be out of date at the time of publication.

Books

What type of book could you write and get published? A specialised monograph comes to mind, a textbook for undergraduates or graduates, a handbook, or popular non-fiction. With a few exceptions, this is a labour of love so you need to consider it very carefully. Remember that, in the sciences, original papers are more likely to get you recognition and to progress your career.

Unless you have been approached directly, before taking the time to write a book, you will need to look for a publisher. The best chance of getting published is to find a company that publishes books on similar subjects – look at several publishers' websites, look in university or departmental libraries and go online. Then make a list of possible publishers and approach them. If a prospective publisher expresses an interest in your work, you will be asked to submit several items, the most important being an outline proposal, your CV and maybe a sample chapter. If, after the internal and external review of your proposal the answer is 'yes', you can start planning. If 'no' then you either approach another publisher or shelve your plans. As requirements vary for each publisher, you will have to check guidelines for authors and may have to alter the form of your submission in order to comply with them. It is best to sign a contract before starting to write a science book.

Having signed, the main consideration for writing a book is finding the time. All of the advice on time management that we have given so far also applies here, but the need for discipline is even greater. You will need to block out a few hours each week and, in addition, look for extended periods of time when you can write undisturbed. This could be during vacation breaks, during sabbatical leave or at weekends – either way, it will have an impact on your science, your family and your social life.

The order in which chapters are written does not matter – once there is a first draft of the book the contents can be reshuffled, some chapters shortened, others lengthened etc. Some people find it easier to do 'broad-brush' work first; others cannot progress unless the details are filled in. So do what works best for you once you have tried different ways. Regardless of your approach, however, revisions are a must. As you go along, give chapters to representatives of your intended audience (work colleagues, students, 'general readers', etc.) to read and critique. Then incorporate their comments as appropriate. The publisher will have the final version reviewed by experts in your field and will decide whether to publish based on their recommendations. You may be asked to make further changes prior to publication.

General considerations

When to write?

This brings us back to Chapter 1 and the weekly planning schedule. Block out 2-hour slots for writing as this will allow you to accomplish something in a day – if things go well you may continue for longer. Treat this time as an appointment with yourself that has to be adhered to. After all, would you want to miss an appointment with your boss? When writing, an 'author' is your boss. And remember to create the best conditions for uninterrupted writing: no emails, no phones and if you share an office, make it clear that for 2 hours you cannot be interrupted. If this does not help, go to a library, work from home if possible or find an empty seminar room and lock yourself in.

> **Pause for thought:** What time of day (or night) are you most efficient at writing? What does your week/month ahead look like? When and where can you block out two precious hours?

When to stop?

Whatever you are writing, it must be read by co-authors and you should discuss/incorporate their comments. Also ask trusted colleagues for further comments and in the case of interdisciplinary papers, ask somebody from another field to read it. This may sound tedious, but it can save you a lot of problems with referees later. Finally, re-check the instructions for authors, comply with them, write a covering letter and send as instructed. Remember that manuscripts are never really finished – they are just submitted – there will always be more that you could add but then you would never publish anything . . .

Is the future totally electronic?

In this chapter we have concentrated on how to write a good paper rather than on different ways of publishing – we are convinced that quality comes first, the method of dissemination comes second. However, we cannot ignore the fast pace of change in science communication methods. Diminishing funding for research is affecting traditional journal publishing as fewer institutions subscribe to journals, online or in paper. Open access journals such as those from the Public Library of Science (PLoS) are becoming more popular and it is inevitable that there will be a mix of communication channels in the future (Russell 2010).

How we did it

In advance of the workshop, participants were asked to read Chapter 4 from Day and Gastel (2006) on what constitutes a scientific paper and also to read examples of published papers. Participants were asked to either look at an interdisciplinary paper 'Inactivation of Viruses' (one of the most downloaded papers in 2007 – http://www.iop.org/EJ/journal/JPhysCM) or to find and compare three papers in the same field in PNAS.

Participants were then asked to analyse their chosen papers in the context of the following guidelines:

We assume that the first step in paper writing should be to assemble the figures and to decide what point each figure makes. The next step is to articulate those points, using one or at most two paragraphs for each figure. Each paragraph should make just a single point. Each paragraph should start with a sentence introducing what is coming in the paragraph and the last sentence should state that point. If this is done properly, the reader should be able to get the message of the paragraph by reading just the first and last sentence of the paragraph. There is no 'recipe' about the order in which the paragraphs should be presented. Generally the paragraphs do not have to be presented in the same order as the experiments were done. One view is that the paper has to tell a good story and the paragraphs should be ordered to narrate that story. However, an alternative view is possible: the most significant result should always be presented first.

Specifically participants were asked to answer the following questions: How well does the chosen paper fulfil these criteria? Can the results paragraphs be understood by just reading the first and last sentence? Are the points clear?

Participants were also asked to bring a copy of work in progress to the workshop. In the workshop, the whole group first compared notes on the published papers. In the second part, participants peer reviewed each other's papers, working in pairs.

Summary

Writing a research paper, a review or a chapter in a book have many things in common – in particular they are all really satisfying if you can clear space to dedicate enough time to them. Realistically, two-hour time slots are the minimum needed for creativity and you will need a lot of those slots to see a writing project through to completion. In terms of career progression, research papers are the most important and so you need to make time for writing a priority when your experimental results warrant publication. A quality research paper tells a good story in a clear and engaging way and is backed up by robust data, a good standard of presentation and a considered treatise of the existing literature.

Selected reading

If you only have time to read one book, make it:
Day, R. A. & Gastel, B. (2006). *How to Write and Publish a Scientific Paper*. Cambridge: Cambridge University Press.

Other texts:
American National Standards Institute (1979). American national standard for the preparation of scientific papers for written or oral presentation. *ANSI*, Z39, 16-1979.
Björk, B. C., Roos, A. & Lauri, M. (2009). Scientific journal publishing: yearly volume and open access availability. *Information Research*, 14(1).
Edwards, D. (2010). Ecology: Life after logging. *Nature*, 466, 668.
Hendrix, R. W. (2009). *Jumbo bacteriophages*. Berlin: Springer-Verlag.
Lagendijk, A. (2008). *Survival Guide for Scientists*. Amsterdam: Amsterdam University Press.
Moon, R. M., Riste, T. & Koehler, W. C. (1969). Polarization analysis of thermal-neutron scattering. *Physical Review*, 181, 920–31.
Nishida, K., Ogawa, H., Matsuba, G., Konishi, T. & Kanaya, T. (2008). A high-resolution small-angle light scattering instrument for soft matter studies. *Journal of Applied Crystallography*, 41, 723–8.
Russell, N. J. (2010). *Communicating Science*. Cambridge: Cambridge University Press.
Starbuck, W. H. (2005). How much better are the most-prestigious journals? The statistics of academic publication. *Organization Science*, 16, 180–200.

4

Handling scientific criticism

As scientists we are all exposed to criticism – it is one of the benefits of working in a discipline where rigour is seen as a community responsibility. However, no-one likes being criticised. Importantly, the more you receive criticism, the easier it becomes to deal with – so it is beneficial to actively seek criticism whenever possible. In this chapter we analyse samples of actual criticism and argue that criticism is delivered in four ways, three of which can be beneficial to you even if they sound damning. The four are objective/ constructive and objective/non-constructive; subjective/constructive and subjective/non-constructive.

The theory

Doing science, reporting discoveries and communicating the importance of our activities exposes us to criticism from various sources. It can be peer review of a paper submitted to a prestigious journal; a referee's report on a grant application; aggressive questioning from a member of the public during an outreach lecture; or a scathing article in the popular press. To be successful as a scientist, you need to handle criticism with confidence and grace, yet nobody likes to be criticised, least of all publicly. Some psychological studies argue that, when we are criticised, our response is of the 'fight or flight' type. It is accompanied by physical feelings of discomfort, anger or aggression, in any ratio depending on our personality, and often our response is not a measured one. If you have ever witnessed two rival scientists hotly arguing after a conference presentation, you know what we are talking about. Some understanding of the nature of a scientist – a subject explored by both sociology and psychology (Mahoney, 1979) – may help put this scenario in perspective. Apparently, we have a very idealistic view of what a scientist should be: objective, rational, open-minded, having superior intelligence, integrity and communality (sharing results with others). This view sets our expectations of behaviour, and if the reality turns out to be different, we are likely to be disappointed. The nature of the scientist delivering the criticism – a human factor – will thus influence both your perception of the criticism and your way of dealing with it.

The practice

Dealing with criticism is a skill that can be developed and as with any lasting skill must be practiced frequently. So seek it – start with people you trust, those who understand your work and your formal or informal mentor. Have you got data that you think would make a *Nature* paper? Put it together and ask somebody who has published there whether they think it is good enough. Accept the advice given; work on writing a paper or collecting more data and have another go, this time asking more people. Ideally, they should be outside your field if your ambition is to inform a broad readership. Having gone through this process, by the time you are ready to submit your paper, you will be reasonably certain of the quality of your work. You will also have practised arguments and counter-arguments and possibly improved your original idea as well. Being proactive in seeking criticism also helps you to detach yourself from feeling personally criticised: it is not that you are stupid, but that your data are not watertight.

Types of criticism

Criticism is delivered in four ways, three of which can be beneficial to you even if they sound damning. The four are objective/constructive, objective/non-constructive, subjective/constructive and subjective/non-constructive.

Objective/constructive and objective/non-constructive

This is, for example, when someone tells you that the data you are presenting do not convincingly support the conclusions you present and gives you a reason why it is not convincing – possibly by giving you some new data or suggesting ways of collecting more appropriate data (see example, in Box 4.1). They can either criticise you because they are trying to help you (constructive) or because they don't like you or your science (non-constructive). Either way, although this can often feel like the most damning criticism, it is the one you should be most grateful for. Often, we convince ourselves of a particular interpretation even before all of the experiments are complete. In so doing, we effectively eliminate the chance of the remaining experiments overturning our interpretation because they are

---

Box 4.1
An example of objective criticism

This is an extract from a manuscript review – the final recommendation was to reconsider after major revisions as it was deemed to contain fundamental results of sufficient novelty and significance:

> The main merit of this paper is to corroborate vibrational data from Inelastic Neutron Scattering, Raman and Infrared spectra for the two polymers polyisobutylene and polyisoprene. The comparison shows that some previous assignments are not in agreement with the neutron data. The authors, however, unfortunately do not contribute much towards a better assignment. This is due to their quite crude data analysis procedure using multiple Gaussians. No attempt is made to use packages which exist to calculate neutron vibrational spectra. Molecular dynamics (MD) simulations have been shown to be extremely valuable in this field as well, and ... ( ) ... both references are not cited.

In response the authors said:

> Firstly, we would like to thank the referee for directing us to two references in the literature that we somehow overlooked. They indeed provide the basis for a more quantitative argument, and as suggested we have compared our measurements with the MD simulations by Alvarez *et al*. To this end, we have calculated the generalised density of states function as shown in new Figs. 4 and 5. The conclusions stemming from this comparison are detailed in the text.
>
> The reference to 'quite crude data analysis procedure' is a little surprising, since it is a widely accepted procedure used to identify peak positions [cf, for example, Braden *et al*., *J. Chem. Phys.* **111**, 429 (1999)]. The only difference between our procedure and that used by Braden *et al*. is that they fitted several features separately, while we fitted all the features simultaneously.
>
> The existing packages to calculate neutron vibrational spectra are not directly applicable to spectra obtained from polymers. A widely accepted reason for this is the complexity of dynamical coupling between different excitations. As far as the package used by Alvarez *et al*. is concerned, it is a commercial, very expensive package to which we have no access nor funds to purchase it. We are therefore unable to simulate different models at present.

This correspondence shows the usefulness of objective and constructive criticism. Having taken these and the other referee's comments into account, the authors were able to improve the paper and it was accepted for publication. Note, however, the way in which the author responded to the comment about 'crude analysis' – other papers were quoted and the differences between approaches spelled out. This was an objective response.

---

designed to prove it. Similarly, we can convince ourselves that a particular dataset is robust enough for publication when it actually is not (and if you were reviewing a paper where someone else presented it, you would say it was not robust enough). Because this type of criticism directly challenges your scientific judgement, it is hard to take. However, most of the time in cases like this, you are wrong and your critic is right – so save yourself a lot of angst and face up to it quickly. If it happens to have been delivered with a non-constructive aim, that negativity is still to your advantage.

> **Pause for thought:** Look at a recent review of one of your papers. Has the referee been objective? How was the criticism addressed, and if the paper was published, was the quality improved?

Subjective/constructive

This happens when someone presents their opinion or interpretation of something that differs from yours. They recognise that your interpretation has merits but prefer their own explanation (see example in Box 4.2). In this case, they tell you because they are trying to resolve the scientific problem. As such, they are being constructive. However, they may still come across as aggressive if, for example, they are pushed for time. Do not interpret this type of criticism as a personal attack. This is someone who disagrees with your scientific interpretation – they are not judging you as a person. Often, this type of criticism helps you explain your case in a clear manner. Incorporating their views into your thought process may well help you work through a problem. Indeed, it is likely that doing this will make your interpretation more widely acceptable.

Box 4.2
Review: An example of subjective/constructive criticism

Other interpretations are possible for the data in Figure 2. The build-up of Mg-protoporphyrin IX and/or the methyl ester could result from a specific defect in the methyltransferase or the cyclase. On the other hand, early precursors such as Mg-protoporphyrin IX and/or the methyl ester build-up during prolonged feeding of ALA to wild type. Therefore, the data in Figure 2 are also consistent with a general reduction in chlorophyll biosynthetic capacity that might result from reduced flux in steps 10–15 (Figure 4) as well as a very specific defect is steps 11 and 12 (Figure 4).

Authors response:

Both Reviewer 3 and Reviewer 4 felt we omitted alternative interpretations of the data presented in Figure 2. We have expanded Figure 2 with TEM images and more spectra. Furthermore, a new supplementary figure has been included to further substantiate our claims. We believe that this also provides good evidence that the genes are required in the dark as well as the light, as questioned by Reviewer 4.

Subjective/non-constructive

This, for example, is when someone tells you that they do not think your work is interesting or that they do not believe your data. You gain nothing by engaging with this type of critic. If the view is expressed in a review and the paper is rejected, it is highly unlikely that the editor used this statement as a basis for a reject. Walk away and forget it, there is nothing constructive you can do about it.

Review process

A formalised version of scientific criticism is the peer review process (Lagendijk, 2008). While discussions with your laboratory colleagues and those further down the corridors are a form of peer review, they are likely to be informal, as is the research seminar given to your group when you invite comments on your work. These methods are open and transparent – everybody sees who is talking and you see who is offering criticism – and people tend to be polite. This is not always the case when you submit a paper or a grant application. In the sciences the peer review process most often takes the form of a *single-blind review* (the identity of the reviewer of the proposal or paper is not revealed). Sometimes, a *double-blind* review is used where the identities of both the applicant and the reviewer are hidden. More recently, *open review* has become an option where both applicants and reviewers know each other's identity, the applicant receives full, signed reviews and authors invite comments on draft publications posted on websites.

For publications, peer review is a powerful mechanism set up to ensure the quality of papers published. In itself it is an object of criticism, especially the single-blind type. At worst, it can lead to delaying or stopping a competitor's paper and stealing priority, or to accusations of nepotism when awarding grants. However, so far there is no better mechanism to ensure that the best science gets supported and published. In this regard it is important to remember that interactions between authors and reviewers are mediated by editors. While the referees give their opinions, the editor decides whether your paper will be published or not (Hames, 2007). With the proliferation of papers and multiplication of sub-areas of research this is a difficult task. It is worth taking time to understand this part of the publishing process before submitting your paper to a given journal.

How we did it

In preparation for this workshop we asked participants to think of some examples of criticism that they had received and to come to the session willing to discuss how they had responded to it. Examples of criticism included referee reports, discussions at workshops and conferences, informal feedback from colleagues etc. In the workshop, the cases were discussed in the context of intent by the critic, perception by the individual and method of response.

The suggested pre-workshop readings were two brief chapters: Chapter 4 in Day and Gastel, 2006 on how to be a reviewer, and Chapter 11 in Lagendijk, 2008 on how to handle reviewers.

Summary

In this chapter we address one of the basic skills needed as a scientist – the ability to receive and handle criticism. The bottom line is that most criticism will improve your science and is thus a positive factor; however, it is often hard to take. As with most things, practice makes perfect (or at least makes things better), so seek criticism on your terms. For example, encourage colleagues in group meetings to actively think about what is wrong with what you are doing; write brief summaries of the research you are doing every six months and ask a couple of people to critique both the science and the writing. Importantly, remember that it is not just you who finds criticism hard to take – be considerate when being critical of others – be objective and constructive.

Selected reading

If you only have time to read one book, make it:
Lagendijk, A. (2008). *Survival Guide for Scientists*. Amsterdam: Amsterdam University Press.

Other texts:
Day, R. A. & Gastel, B. (2006). *How to Write and Publish a Scientific Paper*. Cambridge: Cambridge University Press.
Hames, I. (2007). *Peer Review and Manuscript Management in Scientific Journals: Guidelines for Good Practice*. Oxford: Blackwell.
Mahoney, M. J. (1979). Psychology of the scientist: an evaluative review. *Social Studies of Science*, 9, 349–75.

Web resources:
Research Information Network (2010). Peer Review: A Guide for Researchers. (www.rin. ac.uk/our-work/research/peer-review-guide-researchers).

5

Writing grant applications

Have you ever heard anyone saying that they didn't get their grant funded because the proposal wasn't well written or because it was ill conceived? Probably not – but often it is the reason why funding is refused. This chapter outlines good practice for grant writing based on project management guidelines, and gives a blueprint for what makes a good application. An application for neutron beam time illustrates the points made. References are made to guidelines provided by major funding agencies.

The theory

Your life as a postdoc or an independent research fellow depends critically on obtaining grants from a funding body, such as the Research Councils in the UK, the National Science Foundation in the US, or the European Commission in Europe. Competition for this funding is fierce and, in order to secure money, you or your supervisor will have to submit a research proposal that will be judged by a panel of experts. Whether funds are awarded will depend greatly on how novel your proposal is (your vision), how well planned it is, and how cost effective it is. To succeed, at least in the short term, your novel idea must fit within a certain funding stream and must be viewed positively by your peers. Whilst it is recognised that there are risks inherent to every project, these risks have to be identified and managed appropriately. Furthermore, as the funding agencies are publicly accountable for the grants they give out, you will need to justify how you intend to spend their money.

Before getting carried away and spending a lot of time writing and developing your ideas, you should check your eligibility to apply for a particular award. Depending on your official status, different options will be open to you. For example, if you are on your first postdoc, you can apply for a variety of personal fellowships or for conference travel funds, but would not normally be eligible to act as a principal investigator on a multi-million pound research programme. At the end of this chapter, we list links to various funding agencies and schemes.

In most grant applications, you are outlining a project. As such, your proposal can be framed and developed in the context of good project management practice. Every big undertaking, such as building the pyramids or the Great Wall of China, required both vision and management, but historically these undertakings were not described as 'projects'. The origins of modern project management go back to NASA's Apollo Project. The objective of Apollo was to send man to the moon and it was achieved at a final cost of $25.4 billion (in late 1960s value). Although Apollo was more of a programme, a *collection of projects*; each run by diverse technological, scientific, and administrative units, its success is widely accredited to the way it was managed, with centralised authority over different units with respect to design, technology, science and communication. This type of management strategy was developed in the late 1950s by building on the 'programme management' concept used by industry and the military in the 1940s. Somewhat confusingly, programme and project management are used almost interchangeably in much of the business literature. We take the view that a *programme* is a larger unit constituting several *projects* (Lewis, 2001). A programme could describe a large research group with several people working on related, but different, research questions, or a collection of research groups working collaboratively, each with a project that contributes to the programme.

All projects, no matter how big or small, can be thought of in terms of *scope, time* and *reliable outcome*. These three elements form the basis of project management, a quick introduction to which is provided in a small booklet published by the Institute of Management (Brown, 2002). The central concept of project management is that of an 'eternal triangle' of time, cost and performance (quality). In functional form (Lewis, 2001):

$$C = f(P, T, S)$$

which reads: 'cost C is a function of performance P, time T and scope S'. Performance is essentially a measure of deliverables and, as such, it helps define the project objectives. Scope describes the limits and extent of deliverables.

In schematic representation (Fig. 5.1) P, C and T are the sides of a triangle that envelops the area of the scope (S). Such an arrangement presents constraints in that only three of the four quantities can be fixed. In scientific projects we often start from the scope (the magnitude or totality of the work to be done and the range of expected deliverables) and typically we have to work within a time restriction: think of 1-, 3- or 5-year project grants awarded by funding bodies. This means that realistically we can only fix *either* performance (objectives) *or* cost.

In theory, all projects have the same basic structure and undergo a life cycle with typical phases of *initiation, specification, design, build, installation/implementation, operation and review* (Brown, 2002). Writing a grant application is the most important phase of a project, i.e. the initiation.

Fig. 5.1. The four project constraints. Performance is sometimes replaced by 'quality'.

The practice

Applying project management principles to writing a grant proposal first requires the adoption of Covey's second habit of successful people (begin with the end in mind) (see Chapter 1) (Covey *et al*., 1994). Put simply, until you have a clear idea of what you want to achieve, there can be no project management. Obviously, there are no rules governing creativity and the generation of new ideas. However, once you know your desired end point, you can implement the four main components of the initiation phase of the project (Brown, 2002):

- Setting objectives
- Defining the scope
- Establishing the strategy
- Deriving the work breakdown structure (WBS)

We add one more to this list:

- Determining the required resources (people, equipment and other costs).

In this chapter we deal with the first three subjects; WBS and time-tabling are treated in detail in Chapter 6.

Getting started

One of the most difficult things about grant writing is deciding whether your ideas are 'good enough' to merit funding. As a bare minimum, your proposal must outline an imaginative and coherent programme that exploits your particular expertise in a given field. Carving out such a niche that distinguishes your expertise and interests from those of your postdoc supervisor, is not always easy. Hence it is important to negotiate with your supervisor, preferably at the start of your contract, to free some time to develop your own research that you can take with you at the end of the contract.

Pause for thought: What areas that you are not working on now would you like to explore or develop further?

Setting objectives

There are several good reasons for setting objectives, apart from the fact that virtually every grant application form has a box labelled 'objectives' or directly requests them. Clearly written objectives enable you to focus on the desired results, define and keep track of progress, organise work according to priorities and measure your success. They will ultimately be used to assess your project performance (P). The acronym SMART defines a helpful tool for writing objectives: Specific, Measurable, Achievable, Relevant and Time-bound (see example in Box 5.1).

Box 5.1
Examples of SMART objectives

Specific: To validate computer simulations and predictions using experimental data from synchrotron radiation, X-ray, neutron and light scattering.
Measurable: To improve the efficiency of current photovoltaic cells by 5%.
Achievable: Preliminary results show that technique x works, hence I will use it to study a new substance.
Realistic: A similar but inferior device to our design has been produced by group y.
Time-bound: The testing stage will be finished by 30 June 2010.

Pause for thought: Write down a couple of objectives for your project in a SMART way now.

Having written your objectives, then query them. Are they consistent? Does everybody involved in the project agree with them? Are they aligned with the objectives of the funding body? As consistency is not always possible to achieve, objectives should also be examined for possible trade-offs between time (T), cost (C) and performance (P).

Defining the scope

These are the limits or extent of deliverables of the project. In setting the scope, the questions to be asked are: '*who*, *where*, *when* and *what*?' For example, John Smith from the Department of Outer Space, University of Researchville and Johanna Smit, Astrobiology Group, Laboratory for Extraterrestrial Life. In the first year of the project, John will generate two models for life outside of Earth, one based on carbohydrates and the other on silicones, while Johanna will check the existing database for signs of life on Mars.

Pause for thought: Describe the scope of your recent or planned project in terms of *who*, *where*, *when* and *what*.

Having established the scope you need to turn your attention to constraints (Figure 5.1). This will help you make an initial estimate of the budget and of the internal and external resources needed. Many institutions have dedicated software packages for budgeting. However, departmental administrators, university research services and experienced colleagues can also help you with this step. Money is one of the greatest constraints: once awarded it is almost impossible to get more if the total project cost is underestimated. Time is the next most important constraint – most people underestimate the time needed to complete a project. It may be possible to gain some extra time, for example your Head of Department may be willing to support your project for a month or two if you make a convincing case. However, if you cannot get extra money or gain extra time, inevitably you will deliver less than promised.

Pause for thought: Identify the constraints on your project's scope now.

Next on the list is strategy but it makes sense to consider the *timescale* of the project beforehand. For example, designing a piece of equipment may take half a year, whilst building it could take a further year. All timescales are approximate at this stage, but setting them helps to see where the project is heading. A more detailed schedule can be planned using a Gantt chart once you have your work breakdown schedule (see Chapter 6).

Pause for thought: Estimate how long it would take in your intended project to gather initial data. How long would the main body of work and the data analysis take? How long to write a report, a peer-reviewed paper or to prepare a conference presentation?

Establishing strategy

This is the 'how do you intend to proceed?' question. In a scientific project the main emphasis will be on methodology and will include some detail about experimental techniques, theoretical or statistical methods, etc. as appropriate. There are also relationships with other groups in your department, your university or outside collaborators to consider. This is very important if you need to use a piece of equipment belonging to somebody else or if you can only deliver part of the project. The most difficult part of establishing strategy is setting and clarifying expectations, yours as well as those of others.

Pause for thought: What methodology will be used in your project and who else is likely to be involved?

Risk management

There is a tension between the need to convince project sponsors that the chosen route is likely to be successful and the realisation that the proposed approach may not work. An explicit recognition and management of risk is almost universally required by the funding bodies. Hence this is the point to ask yourself 'what if' type of questions. For example, what will you do if a named researcher leaves the project midway because he/she gets another job? What will be the impact if you fail to generate transgenic lines within the first year of the project? What if your postdoc cannot reproduce vital results because the published method was incorrect or invalid? In each case you will have to consider alternative ways to achieve your objectives. For example, if you intend to model the structure and dynamics of DNA using a particular method and somebody else has beaten you to it, you could consider using a different method or studying another type of molecule.

Pause for thought: What are the risks inherent in your project?

Making an application

Application form

In reality, all of the points made above are hidden in a form that you have to fill in to apply for a grant. While details vary between funding bodies, typically you will be asked for an abstract, objectives, background information, work to be carried out, impact, beneficiaries and cost. Pay special attention to explaining the broad sig-nificance of the proposed research – if there is no dedicated box on the form, put a section at the end of your case for support. Finally, fill in the proposal form very carefully keeping to the rules of the funding agency, particularly with regard to font size and page limits.

Case for support

The precise format of the case for support will vary, depending on the required length of the proposal, the nature of the science and the funding agency (one example is shown in Box 5.2). However, a few general 'rules' apply.

- The title should be as short as possible, but must convey the aims.
- The abstract must clearly convey the research question(s) to be addressed, the work to be carried out and the likely significance of any results obtained.
- The background information must convey the conceptual framework and indicate why the area is interesting. It must make the subject sound exciting even to non-specialist readers.
- The aims must state the research question to be addressed and indicate how it will be addressed.
- The objectives must outline the targets that are deemed to be deliverable.
- Preliminary data should be placed together in a section that provides unpublished information that leads into the proposed work.
- The programme of work should be structured such that each section aligns with an objective. Logistical considerations (including risk management) should be outlined, either in each section or as a section at the end of the programme of work.
- The time-table is often best demonstrated graphically so that it is easy to see the workload in each year. Many funding agencies expect or ask for a Gantt chart (see Chapter 6).
- There should be a short section at the end of the proposal outlining the significance of the work in the context of the original aims.
- The impact of your work must be clearly stated.

Having drafted the case for support, get a non-specialist to critically read the abstract and background. Incorporate suggestions and then have another non-specialist read the revised version.

Box 5.2
Example of a successful application for neutron beam time at a central facility

Here we describe how a successful application fits within the framework described above. In this case the proposal has a two-page limit and specific constraints on what has to be included.

Proposed experiment

Name: Dr Mark Telling   Experiment Title: QENS Study of Molecular
                                                                    and Intra-Molecular
                                                                    Reorientation in the
                                                                    HAB Liquid Crystal

Abstract

A brief summary of the proposal, set into the context of your research programme, including key objectives.

   We propose to study the complex molecular motions for the series of the protonated and partly deuterated samples of a liquid crystal HAB as a function of temperature in different phases.

   Objectives:

- To separate the rotational motion of whole molecules around the long axis and rotations of particular molecular fragments (tails, benzene rings).
- To compare the obtained values of correlation times with those obtained from dielectric spectroscopy and ^{2}H NMR techniques.

Experiment description

Background

Molecular motions existing in liquid crystals are of interest to both academia and industry. Liquid crystals often exhibit a rich polymorphism, and various phases differ not only with respect to the arrangement of molecules, but also with respect to the reorientational (stochastic) motions of both whole molecules and their parts (moieties). Detailed knowledge of such motions is needed in search for applications such as fast switching displays. [1] Liquid crystal *p-di-n*-heptyl-azoxybenzene (HAB) is interesting and well suited for QENS study because its molecule is almost ideally symmetrical with respect to its short axis, thereby decreasing the number of possible contributions to the QENS broadening (Fig. 5.2.1). (In other liquid crystals of similar 'architecture' but with chains of different length, large number of such components often renders a QENS measurement rather inconclusive.) Moreover, it can be synthesised with its rings or aliphatic chains deuterated (see Fig. 5.2.1). On cooling, HAB exhibits the following phase transitions: Isotropic – 69.6 °C – Nematic –53.5 ° C – Smectic A –26.1°C – Crystalline.

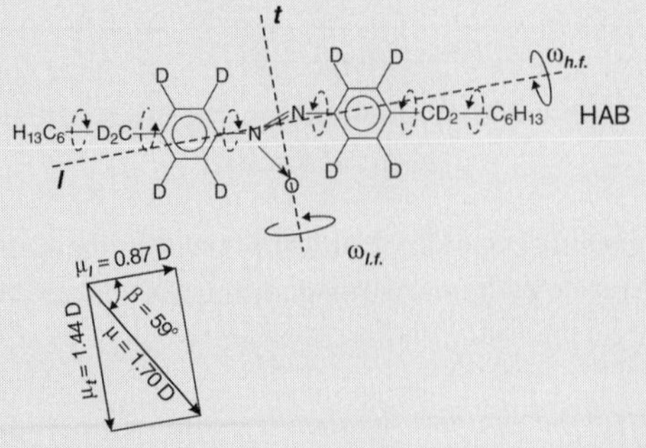

Fig. 5.2.1. Structure of the HAB-D$_{12}$ (ring-deuterated) molecule. The lowest (l) and highest (t) inertia moment axes are shown with the respective

whole-molecule rotations observed in the dielectric studies. The intramolecular
rotations about the N–C and C–C bonds are marked by dashed lines.

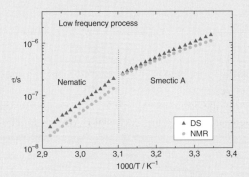

Fig. 5.2.2. Comparison of the l.f. relaxation times determined from the DS and
NMR results.

Recently, some of us [2] performed dielectric spectroscopy and NMR studies of HAB in
order to determine the correlation times characterising the reorientation motions of the
molecule as a whole and/or its particular fragments (see Fig. 5.2.2). Dielectric spectroscopy
allows to detect both principal rotational motions: low frequency (l.f.) connected with the
rotations around the short axis, and high frequency (h.f.) characterising the motions around
the long axis. The NMR studies yielded the rotational diffusion coefficients $D_\parallel$ and $D_\perp$ that
characterise the tumbling, spinning and internal motions of the two benzene rings. In fact, a
relation between the correlation times $\tau^L_{m,n}$, which are related to dielectric relaxation times,
and the overall molecular rotational diffusional constants ($D_\parallel$ and $D_\perp$) that can be obtained
from NMR relaxation data, could be derived from theoretical studies on uniaxial mesophases.
However, a quantitative comparison between dielectric and NMR results is possible only in
the case of the experimental dielectric low-frequency relaxation time ($\tau_{l.f.}$), which can be
identified with the correlation time $\tau^1_{0,0}$ and therefore related to the diffusional coefficient $D_\perp$
obtained from NMR. On the contrary, the experimental dielectric high-frequency relaxation
time ($\tau_{h.f.}$) is a combination of different relaxation components ($\tau^1_{1,0}$, $\tau^1_{1,1}$ and $\tau^1_{0,1}$) and can be
only qualitatively compared with the rotational diffusional constant $D_\parallel$ (Fig. 5.2.2).

There are theoretical models allowing for calculation the correlation times in the N
phase for the l. f. process only. The l.f. relaxation time $\tau_{l.f.}$ is related to $D_\perp$ and the order
parameter $S = <P_2(\cos\theta)>$. The obtained agreement seems to be satisfactory in this case.
However, the other correlation times cannot be interpreted at present due to a lack of
knowledge of relationships between them. Neutron scattering measurement is the only
way to disentangle rotational and orientational motions in this crystal.

Work proposed

We propose to study a series of selectively deuterated HAB molecules in order to
determine the types of reorientation and the corresponding characteristic times that
HAB and/or its moieties ('halves') undergo in various phases. Many years of experience
with reorientation of liquid crystals indicate that, in order to understand the nature of
such motions, one has to compare results from various complementary techniques

since each of them looks, in fact, at a slightly different phenomenon. For example, dielectric

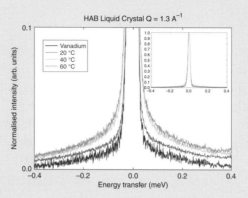

Fig. 5.2.3. HAB test run on OSIRIS. Fully hydrogenated sample in crystalline, smectic and nematic phases.

relaxation yields time characteristics of reorientation of the dipole moment only. This is the rationale behind this experiment. We have already carried out a very quick test measurement on OSIRIS, and demonstrated (Fig. 5.2.3) the feasibility of such an experiment.

In order to collect publishable quality spectra, we need to count at least 12 hours per spectrum. Taking into account two samples, one deuterated and one non-deuterated, three temperatures (one in each phase) and calibration runs, we request five days of beam time on OSIRIS.

References:

[1] see, e.g. http://moebius.physik.tu-berlin.de/lc/lcs.html

[2] Dynamics of 4,4′-di-n-heptyl-azoxybenzene (HAB) by means of dielectric and ^{2}H NMR relaxation measurements, V. Domenici, J. Czub, M. Geppi, B. Gestblom, S. Urban and C. A. Veracini, in preparation.

List publications arising from recent ISIS experiments:

- 'Evidence of localized fast picosecond process in glassy poly(methyl methacrylate) far below Tg', T. Kanaya, I. Tsukushi, K. Kaji, B. J. Gabrys and S. M. Bennigton, *Phys. Rev. B*, **64**, 144–202 (2001).
- 'Neutron Compton scattering studies of stretched polyethylene', B. J. Gabrys, W. Zajac, J. Meyers and M. S. Kalhoro, *Applied Physics A*, **74**[Suppl.], S1645-S1647 (2002).
- 'Reflectivity studies of ionomer blends', B. J. Gabrys, D. G Bucknall, A. A. Bhutto, R. Braiewa, D.Vesely and R. A. Weiss, *Applied Physics A*, **74**[Suppl.], S336-S338 (2002).

Has this or a similar proposal been submitted to another facility recently? Yes ☐ No ☒
If yes, please give details:

> **Our analysis::**
> **Abstract**:   The abstract clearly conveys the research question to be addressed. In this case key objectives had to be included. Objectives should always outline the targets that are deemed to be deliverable.
> **Background**:   The background information conveys the conceptual framework and indicates why the area is interesting. In this example it also includes preliminary data that leads into the proposed work.
> **Programme**:   The programme of work is brief, but includes both logistical considerations and a timescale of experiments.

This work was subsequently published in: Zajac W, Urban S, Domenici V, Geppi M, Alberto Verancini C, Telling MTF, and Gabrys BJ (2006). Stochastic molecular motions in the nematic, smectic-A, and solid phases of p, p'-di-n-heptyl-azoxybenzene as seen by quasielastic neutron scattering and ^{13}C cross-polarization magic-angle-spinning NMR. *Physical Review E*, **73**, 051704.

Things to remember

- Make sure the proposal is 'tidy' – remove typos and general sloppiness.
- Keep the proposal focused and the workload manageable. If you are not going to be doing the work yourself, it may go much more slowly than you imagine.
- The proposal may be reviewed by someone in a very different field from you – make sure the take-home message will be clear to them even if they don't understand the detail.
- Don't put too much experimental detail in – it detracts from the big picture.
- The best-case scenario when you submit a grant proposal is that a third of the work is already under way, a third will (definitely) get done and the final third may get done if all goes well (blue skies research). Under these circumstances, you will be able to guarantee a successful end of grant report. That, in itself, will go a long way to helping the next grant get funded.

The most common errors

- The proposal is far too ambitious – normally with respect to the amount of work proposed, but sometimes because there is insufficient preliminary evidence to validate the proposed approach.
- Priorities are not clear.
- Timetable is unrealistic.

- Insufficient effort is made to convey why the work should be of interest to a wide range of people.

Other tips

- Keep up with developments in your field by scanning the literature and attending conferences.
- If possible, get a copy of a successful proposal close to your field.
- Participate in workshops organised by funding agencies.

How we did it

We have experimented with the delivery of this workshop over the years. In one version, participants are asked to arrive, having thought about a project that they would like to pursue. In the workshop they then draft each of the five main components of the initial phase of a project (objectives, scope, strategy, timeframe, resources). After each of the five steps, the participants discuss what they have written in pairs or small groups. In this way, each participant leaves the workshop with an outline of a grant proposal. In another version, proposals are written before the workshop and are distributed ahead of time to groups of three. During the workshop, participants work in those groups to critique and refine each other's proposals. Note that, in this case, groups of three work much better than pairs because it allows different views of the same proposal to be expressed.

Summary

This chapter gives a short guide to writing grant applications in the framework of project management. Understanding the constraints of scope, time, cost and performance should help to design a manageable project and writing SMART objectives will focus your mind on project deliverables. The key to a successful application is then a clearly written case for support. Your project should sound novel, significant and exciting even to non-specialists; your strategy should be logical with risks accounted for; and your resources and timeframe should be sufficient to deliver your objectives.

Selected reading

If you have time to read only one book, make it:
Brown, M. (2002). *Project Management in a Week*. London: Hodder & Stoughton.

Other texts:
Covey, S. R., Merrill, A. R. & Merrill, R. R. (1994). *First Things First*. London: Simon & Schuster.

Lewis, J. P. (2001). *Project Planning, Scheduling, and Control: A Hands-on Guide to Bringing Projects in on Time and on Budget.* London: McGraw-Hill.

Web resources:
A Guide to writing SMART objectives: http://www2.eastwestcenter.org/research/popcomm/pdf/6_Elements_of_a_Communication_Strategy/Smart_objectives.pdf
Guide to grant writing: http://www.hfsp.org/how/ArtOfGrants.htm
NASA Apollo project: http://history.nasa.gov/Apollomon/Apollo.html

Links to Funding Agencies (all last accessed 6/12/10).

In the UK
Research Councils UK
Arts & Humanities Research Council (AHRC) http://www.ahrc.ac.uk/Pages/default.aspx
Engineering and Physical Sciences Research Council (EPSRC) http://www.epsrc.ac.uk/Pages/default.aspx
Biotechnology and Biological Sciences Research Council (BBSRC) http://www.bbsrc.ac.uk/home/home.aspx
Economic & Social Research Council (ESRC) http://www.esrc.ac.uk/ESRCInfoCentre/index.aspx
Medical Research Council (MRC) http://www.mrc.ac.uk/index.htm
Natural Environment Research Council (NERC) http://www.nerc.ac.uk/
Science and Technology Facilities Council (STFC) http://www.stfc.ac.uk/

Other bodies
The Royal Society – http://www.royalsociety.org/
Wellcome Trust – http://www.wellcome.ac.uk/
Leverhulme Trust – http://www.leverhulme.ac.uk/
Human Frontier Science Program – http://www.hfsp.org/
National Institute for Health Research (NIHR) – www.nihr.ac.uk

Europe
Gateway to European Research & Development – CORDIS – http://cordis.europa.eu/home_en.html
European Research Council (ERC) – http://erc.europa.eu/

US
National Science Foundation – http://www.nsf.gov/
National Institutes of Health – http://grants.nih.gov/grants/oer.htm
Howard Hughes Medical Institute – http://www.hhmi.org/
Department of Energy – http://www.energy.gov/

The above list is not exhaustive, and your Research Services Office will advise on opportunities, eligibility and calls for submission. Information about personal fellowships is included in Chapter 9.

6

Tools for managing research projects

Successful scientists have to be able to organise research projects. Scientific projects can be efficiently organised using tools developed for management practice, at the small cost of seeing beyond the language. It is vital to clearly set out the various stages of a project, and developing a work breakdown structure (WBS) is often the first stage. A well written WBS can then be used to produce Gantt and PERT charts to monitor project progress. This approach is illustrated here using proprietary software (Microsoft Excel and Microsoft Project) and examples from materials science.

The theory

All projects have a beginning, a middle and an end, with stages that can be classified as in Fig. 6.1.

Notably, project outcomes are never written in stone – they are anticipated at the start of the project but are often modified as the project progresses. This is accepted as an inherent property of every project, a property that emphasises the need for flexibility at the detailed planning stage, and for monitoring and control throughout. This is true both for scientific and generic projects.

There are several definitions of the word project – we like one by J. M. Juran: 'A project is a problem scheduled for solution'. As the essence of research is posing and solving problems, this definition would be unremarkable if not for the word *scheduled*. This one word brings a problem from the domain of possibility into the realm of action. You may already be scheduling aspects of your research project in your weekly calendar. However, here we need to look at the bigger picture – from the moment a project is conceived to its timely and successful end.

In Chapter 5 we discussed project initiation. We now need to look at project organisation and more detailed planning. This means we need to find a systematic approach to answer the following questions:

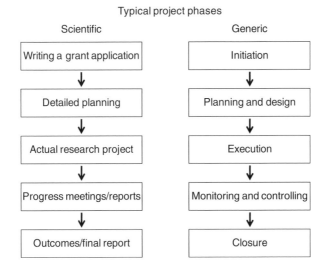

Fig. 6.1. Comparison of typical scientific project phases with generic project phases.

1. What must be done?
2. Who will do each task?
3. How long will each task take?
4. What materials, supplies and equipment are required?
5. How much will each task cost?

A Work Breakdown Structure (WBS) allows you to answer the 'what' from the above list. It helps to break down a complex project into a number of tasks which need to be done and should ensure that you have not forgotten anything important until it is too late. While WBS does not involve putting tasks into sequence, it brings the whole project together, and serves as a basis for answering Questions 2–5. The schematic representation of a WBS also makes it easy to discuss with others. A popular analogy here is 'the sliced salami' – when cut into fine slices, salami is appetising; a WBS makes the project appear manageable.

The practice

A simple example

Imagine that your project is to set up a laboratory experiment for physics students to confirm Ohm's law. The empirical Ohm's law states that a current flowing through a circuit containing a conductor is directly proportional to the potential difference across this conductor. So your WBS for this project may look like Fig. 6.2.

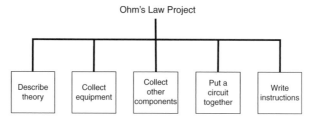

Fig. 6.2. Top level Work Breakdown Structure for an experiment to confirm Ohm's law.

Now you can talk it through with your more experienced colleagues, and perhaps refine 'describe theory' to consider limitations, for example, this law does not apply to all conductors.

But is the WBS really useful for research projects? Yes it is – while you cannot plan for discovery you can make use of conditional branches (Lewis, 2001). So, you devise a plan and depending on the outcome of your experiments you will follow a particular direction.

Pause for thought: Sketch a WBS for a research project that you are currently working on.

With the WBS in place, you can tackle Questions 2–5. How much will it cost? How many experiments will be conducted at the same time? Are there sufficient voltmeters, amperometers and computers available? Is there any technical help available or will you have to carry out all of the tasks yourself? If there are more people involved in the project, you need to assign them different tasks. It is also helpful at this stage to consider any intermediate results of a project – the *milestones*. The achievement of milestones can then be given time estimates and be presented in the form of a Gantt chart (Fig. 6.3).

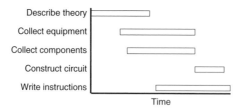

Fig. 6.3. A Gantt chart for Ohm's law project.

The Gantt chart is an essential tool to schedule, monitor and control a project. Its form is simple: on the *y*-axis there is information about tasks that comprise the project and on the *x*-axis the time period required for completion of each of these tasks is represented. There are a wide variety of acceptable representations, ranging from a table drawn in a word-processing package through to a table produced in a general software package such as Excel, to quite complex graphs obtained using dedicated software such as Microsoft Project.

The chart in Fig. 6.3 needs time calibration – depending on the size of the project, it can be hours, days or months. However, even with a timescale, it does not tell us about the relationship between tasks, and especially whether the completion of any one task depends critically on others. For many projects, it is sufficient to schedule tasks in a logical order but for large projects more sophisticated management tools are needed (Lewis, 2001).

> **Pause for thought:** Sketch a Gantt chart for your project now. It is essential to draw it first by hand. Once you are happy with your chart, you can use Microsoft Excel for a more polished touch.

A complex research project

Three-year research projects are more complex than the Ohm's Law project outlined above – so how does such a project map on to the methods discussed? Below we develop a detailed project plan for the example shown in Box 6.1.

Box 6.1
Example of a 3-year research proposal

Studies of biomacromolecules using a combination of neutron scattering and molecular modelling.

   Project aims:

1. To model the interaction between cationic polymers (polysaccharides) and DNA, and of polysaccharide nanoparticles with the cell membrane at both atomistic and molecular levels.
2. To focus on *coarse-grain methods* which describe time domains ranging from several nano- to microseconds, and operate on a spatial domain of 1 to 100 nm.
3. To validate simulations and predictions using experimental data primarily from synchrotron radiation, X-ray, neutron and light scattering.

Development of WBS

To develop the WBS, we first ask what needs to be done – it helps to sketch out thoughts as shown below.

   Need to do:

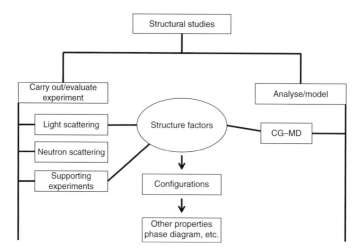

Fig. 6.4. A WBS for structural studies of biomacromolecules using a combination of neutron scattering and molecular dynamics. CG – coarse grain, MD – molecular dynamics.

1. Preliminary analysis of neutron scattering and synchrotron data – obtain structure factors.
2. Analysis of light scattering data.
3. Mapping out the phase diagram of polysaccharides.

Figure 6.4 shows schematically the connection between structural studies (left hand side) and simulations (right hand side). Structure factors are the most important quantities linking results of experiments and simulations; hence they occupy a central position in the diagram. They will be directly found from neutron and light scattering experiments. On the other hand, they will be simulated in coarse grain molecular dynamics (CG – MD) 'experiments'. Subsequently, molecular configurations will be obtained, and in the next step other properties will be generated.

Task allocation

Having established what needs to be done, we can now look at the allocation of tasks and the share of responsibility between people involved. This step is crucial to the success of the project. One person has to have overall responsibility for the whole project and he or she must check overall progress and make sure that the team communicates. On large projects, such as those in particle physics, a professional manager is frequently appointed. Box 6.2 shows task allocations for the project.

Box 6.2
Example of task allocation

Named researcher: 80% modelling, 20% experimental, day-to-day project management and writing
    Senior specialist: 2 hrs/week on model development of permeation
    PI 1: overall project management, guidance in MD simulations
    PI 2: sample preparation, neutron scattering and synchrotron experiments
    Collaborator 1: help with experiments, analysis of synchrotron and neutron scattering data using Niebuhr's programs
    Collaborator 2: light scattering data gathering and preliminary analysis

---

Project outline and milestones

So, once you know who is going to do what, the next thing to do is to work out the desired order of events and identify when certain milestones will be reached (see Box 6.3).

Box 6.3
Example of project outline and milestones

Year 1

Molecular dynamics (MD) code adaptation

1. Propose model of structure–functionality link in terms of multi-scale modelling.
2. Data acquisition (apply for neutron and synchrotron beam time) and preliminary interpretation as input for model development.
3. Refinement of preliminary structural phase diagram for polysaccharides: extension of polyelectrolyte theory to semi-dilute solutions.
4. Interpret relevant parameters such as r_g and the correlation length ξ in terms of scaling theory of polyelectrolytes.
5. Build coarse-grain model of a known molecule (literature or own data).
6. Write a short report on building this model.
7. Import neutron scattering data from dilute solutions of chitosan into MD programmes.
8. Analysis of synchrotron data from dilute solutions of chitosan using Niebuhr's programmes.
9. Extract structure factors.
10. Propose configurations.

Milestones

Progress meetings (every 6 months), report/publications, structure factors (measured and simulated) for chitosan, conformations in dilute solutions, quantitative information about interactions between molecules themselves as a function of environment.

Year 2

Code validation and adaptation to reference polysaccharides and membranes

11. Coarse Grain (CG)-modelling of chitosan.
12. Input of resulting structure factors/configurations into MD.
13. Atomistic refinement.
14. CG-modelling of model membranes.
15. Atomistic refinement of model membranes.
16. Study of absorption/desorption of polyelectrolytes at charged surfaces: experiment and CG-modelling.
17. Comparison of simulated and measured structure factors.
18. Refinement of CG-configurations by MD simulations.
19. Repeating the process (12–17) for heparin.
20. Theoretical study of permeation through membranes aided by MD simulations.

Milestones

Progress meetings (every 6 months), report/publications, CG-models of chitosan and heparin.

Year 3

Advanced studies: prediction and proposing new experiments

21. Bilayers: simulation and preliminary experiments.
22. CG model of DNA + specific membrane.
23. Analysis of existing DNA + membrane data.
24. Model DNA structures within the framework of phase diagrams.
25. Atomistic refinement of the above.
26. Prediction: design new experiments on DNA with membranes.

Milestones

Progress meetings (every 6 months), report/publications, structural model of
DNA + membrane under different experimental conditions, development of a model for big biomolecules permeating through lipid membranes.

Dependencies and risks

This project is not yet ready for scheduling as we have to spell out the dependencies – which tasks have to be accomplished first before the next ones can be tackled. There are some obvious ones in this project – an experiment has to be successfully conducted before the data can be analysed. Sometimes you have to know the outcome of an experiment in order to conduct another one; sometimes experiments can be carried out concurrently. The former brings more vulnerability to the project whereas the latter decreases the risk of failure. In this context, you also need to identify potential risks and make sure that you have contingency plans in place (see Box 6.4).

Box 6.4
Example of risks and contingency plans

1. No student available for sample preparation – serious – get another student to do it.
2. No beam time allocated first time round – serious – reapply, start from light scattering data, start from extension of Dobrynin & Rubinstein theory.
3. Experiments fail – serious – repeat, change substance.
4. Named researcher leaves the project – catastrophic – employ a postdoc and/or change the scope of the project.
5. Underestimation of computation time – serious – get more computing resources or diminish the size of a system.

Pause for thought: Analyse one of your research projects as outlined above.

The polished Gantt chart for the above project is shown in Fig. 6.5.

Scope creep

As projects progress, it is important to watch out for 'scope creep'. Scope creep causes a project to overrun its time and/or cost. Often, it may be hard to spot – a stakeholder asks you to bring in a small change or to add an extra feature – and before you know it you haven't accomplished your major objective. For example, adding one more chapter to this book would only be a 5% change in terms of the total number of planned chapters. Depending on the subject of this additional chapter, however, it could take anything from 2 weeks to 2 months to write, which could result in a failure to meet the publisher's deadline. Scope changes can be controlled by adjusting constraints, but less time available may result in poorer quality outputs and more time will incur higher costs or require more resources.

How we did it

The material presented in this chapter was covered in the workshop on writing grant applications (outlined in Chapter 5). Participants were asked to sketch a rough Gantt chart for their planned project.

Summary

We have provided a short guide to useful tools for project management. A WBS is the starting point as it gives an overall view of tasks to be completed. However, the

ID	❶	Task Name	Duration	Start
1		Modelling supramolecules & membranes	186.4 wks	01 Nov '06
2		Preliminary experimental investigations:	20 wks	01 Nov '06
8		Milestone: determination of the structural phase diagram:	0 wks	20 Mar '07
9		Report: electrostatic interactions	4 wks	21 Mar '07
10		Year 1:Method-MD and RMC code adaptation	52 wks	01 Jun '07
11		Develop CG model	24 wks	01 Jun '07
18		Progress meeting 1	1 wk	16 Nov '07
19		CG model available	0 wks	22 Nov '07
20		Study Structure factors	26 wks	23 Nov '07
26		progress meeting 2	1 wk	23 Nov '08
27		Structure factors (measured and simulated) for chit	0 wks	29 May '08
28		Qunaitative info about interactions	0 wks	29 May '08
29		theoretical study of permeation	48 wks	01 Jun '07
33		diffusion constant D	0 wks	01 May '08
34		Year 2 Code validation and adaptation to reference	48 wks	30 May '08
35		CG-modlleing of chitosan	28 wks	30 May '08
44		Progress meeting 3	1 wk	12 Dec '08
45		CG-models of chitosan;	0 wks	18 Dec '08
46		Study heparin	14 wks	19 Dec '08
50		progress meeting 4	1 wk	27 Mar '09
51		CG-models of chitosan and heparin;	0 wks	02 Apr '09
52		Report CG forcefield (MD) with the RMC ready	0 wks	02 Apr '09
53		theoretical study of permeation	48 wks	30 May '08
57		sorption coefficient S	0 wks	30 Apr '09
58		Year 3 Advanced studies: prediction and proposing	56 wks	01 May '09
59		Membranes	48 wks	01 May '09
65		Progress meeting 5	1 wk	11 Dec '09
66		Report structural model of DNA + membrane ready	0 wks	17 Dec '09
67		theoretical study of permeation	48 wks	01 May '09
71		permeability P	0 wks	01 Apr '10
72		Refinements and final reports	23 wks	18 Dec '09
77		Progress meeting 6	1 wk	18 Dec '09
78		Final report available	0 wks	27 May '10

Fig. 6.5. Example of a Gantt chart for a 3-year research project.

Gantt chart is the major tool used to schedule and monitor project plans. This can easily be generated in a word processing package such as Microsoft Word or in a spreadsheet such as Excel.

Selected reading

If you only have time to read one book, make it:
Brown, M. (2002). *Successful Project Management in a Week*. London: Hodder & Stoughton.

Other texts:
Lewis, J.P. (2001). *Project Planning, Scheduling, and Control: A Hands-on Guide to Bringing Projects in on Time and on Budget*. London: McGraw-Hill.

7

Is there life beyond academia?

In most HE institutions, a lectureship/assistant professorship, whether permanent or fixed-term, is the first step on the career ladder to a professorship. Before applying for such a position, it is therefore important to ask yourself whether an academic job is what you really want. What skills have you acquired that would ensure success outside academia? What else is out there that can be challenging and rewarding? Are you better suited to eventually managing a multi-million-pound corporation or running a charity? Would you be happier as a senior partner in a patent office, as head of research in a governmental institution or as a senior administrator in a university? We explore different options and give examples of very successful people who left academia to make their mark elsewhere.

The theory

It is obvious that the number of people with higher degrees far exceed the number of places available in academia. It is also obvious that academia is not the best career choice for everyone. For example, if you are a great researcher but do not enjoy teaching, then academia is not the place for you. However, your scientific skills – such as analytical reasoning, ability to solve problems in an unorthodox way (thinking outside the box), to communicate to a wide audience and to synthesise diverse sources of information – open many avenues for you.

Before deciding on a career path you will need to think carefully about your principles and priorities (see Chapter 1) and to gather information from several sources. There are many alternative careers in science and most universities organise careers meetings where non-academics talk to postdocs about their own career trajectories following a science PhD and postdoc. Even better, if you know a scientist working outside academia – talk to him or her – what have they done since being a postdoc or PhD student, what excited and inspired them and what frustrated them? You will also need to assess your own skills – most universities have a Careers Service or Human Resources office that can help you do this.

Another key thing to consider is the purpose or mission of your prospective employing institution. Your mission has to fit in, be aligned with and aspire to the same goals as the institutional mission. A good mission statement should demonstrate the institution's values and its commitments to its employees. It should address five key areas:

- *What* is the purpose of the organisation?
- *Who* benefits from the organisation?
- *Why* does the organisation need to exist?
- *Where* does the organisation operate?
- *How* does the organisation fulfil its purpose?

We have a look at some mission statements (sometimes referred to as institutional strategies) in the context of the personal stories that are told below.

The practice

Leaving academia – different career paths

We asked several successful non-academics to tell their story with particular reference to the following questions:

1. Why did you originally decide to study science?
2. What did you like and dislike about scientific research?
3. What skill set did you gain as a research scientist?
4. Why did you decide to leave academia?
5. What was your career path from academia to your current job?
6. What is your current job?
7. Which skills are particularly relevant to your current job?

> **Pause for thought:** Look at questions 1–3 and write down your answers now.

If you like academia and working with academics, but have had your fill of research work, there are several opportunities at universities. More and more senior administrators have PhD degrees and there are also positions in which you support colleagues through help and advice on intellectual property rights. Dr **Debbie Alexander** is Licensing Associate at the Office of Technology Management, University of California San Francisco (UCSF). Her responsibilities include technology management, licensing and outgoing material transfer agreements. The

UCSF Office of Technology Management was established in 1996 with the charge to bring the results of the research and educational programmes at UCSF forward for public use and benefit, with any net revenues derived from licensing those results to industry and/or end-users to be applied for the purposes of supporting the basic research, clinical research and education missions of UCSF (http://otm.ucsf.edu/docs/otmTechMan.asp).

Here is her story:

> The best piece of advice given to me when I wanted to make a career change from academia was 'the perfect job is one that (a) requires skills you excel at AND enjoy and (b) provides opportunities to learn new things'. This magic formula certainly worked for me. I chose a career that integrates my favourite aspects of being a scientist with the opportunity to learn about new fields and I can honestly say that I love my job as a Licensing Associate in UCSF's Office of Technology Management.

> The main goal of my job is to transfer scientific innovations from UCSF to biotech and pharmaceutical companies in order to bring new medicines, medical devices etc. to the market. I manage around 100 inventions at all stages of development and, on any given day, will work on approximately ten of these inventions. Invention management involves ongoing communication with the scientist about their new invention, evaluating inventions for commercial potential, liaising with patent attorneys to manage the patenting of commercially attractive inventions, marketing inventions to relevant companies and finally negotiating licensing deals with companies to develop commercial products. There is certainly a lot of variety and endless opportunities to learn about new areas of medicine, business strategies, and patent law.

> In addition to the job variety and learning opportunities, I gain a lot of satisfaction in knowing that many of the skills necessary for my job came from my career as a scientific researcher. In fact, these skills are so valued in my office, that all of the licensing staff have PhDs. These skills include a broad knowledge of scientific concepts and techniques that can be applied to any field: excellent planning, organisation, multitasking, and problem-solving skills, the ability to quickly grasp and summarise complex information and knowledge of how to successfully work with academic scientists. The latter may sound trivial, but as it turns out, it is by far the most difficult part of my job!

> While Technology Transfer is currently my perfect job, originally, the perfect job for me was as a scientist. I was always top of the class in science subjects at school and was endlessly curious about how living things 'worked'. I also really enjoy travel, so saw science as a great way of working anywhere I wanted in the world (many scientists undervalue this aspect of their career, yet the experience of working internationally and the exposure to scientists with different working methods and ideas, is invaluable). I started off closer to home, with a PhD in the Department of Plant Sciences at Oxford, but then decided to move to Stanford University in California for a postdoc. Not only is there a high concentration of excellent plant science labs in California, but I wanted to live somewhere sunny!

> For the first couple of years I really enjoyed being a postdoc; the work was interesting, my colleagues were stimulating, and outside of the lab I had a great social life, with most

weekends spent outdoors, either hiking, snowboarding or exploring San Francisco. However, I began to realise that while I loved talking, learning and thinking about science, I was bored with bench work. I tried learning new techniques and collaborating with other labs on projects, but I was still bored. I thought of applying for academic jobs (since I would have postdocs and graduate students to do the experiments, that should be better, right?), but realised that my problem ran deeper than just bench work. I wanted to be involved in many different aspects of science, not just what was happening in one lab.

What career should I choose though? I had always assumed that I would run an academic lab, so was completely unprepared for having to rethink my career. I didn't even know what types of jobs were available. Taking my friend's advice, I came up with a list of my skills and thought hard about which ones I wanted to use everyday. I also added in some skills that I wanted to learn or improve on such as business communication, negotiation skills and teamwork. Armed with my list, I attended networking events and asked as many people as possible which career would be compatible with these skills. The two careers people suggested were Technology Transfer and Science Writing.

Knowing very little about either of these careers, I set up a number of informal interviews with people working in these fields and began applying for jobs. When applying for jobs in which you have no experience my advice is to:

1. Learn how to write a skills-based resume and compose a compelling cover letter describing the skills you will bring to the job.
2. Ignore the 'qualifications and experience needed' sections of job advertisements. If the job sounds interesting and you think you can do it, apply.
3. Consider taking some classes, or doing an internship to develop your skills and experience.

Steps 1 and 2 worked very well for me with regard to a career in Scientific Writing. In fact, during a job interview, I was told that I had none of the qualifications or experience they needed, but they liked my covering letter so much that they wanted to interview me anyway! However, by the time I'd done a few interviews, I realised that I was much more interested in a career in Technology Transfer, so I didn't end up taking any of the jobs offered. Unfortunately, I had reached a dead-end with regard to Technology Transfer positions. Most offices simply would not consider applicants without experience of working in a company. Therefore, I turned to step 3 – I contacted some local Technology Transfer offices and asked if they would take me on as a volunteer. I struck gold – UCSF had an unadvertised, informal internship programme already in place, specifically for PhDs. The catch was that you are required to intern at least 10 hours a week, unpaid, and it could be up to a year before a job opened up. I applied immediately and was accepted into the programme.

So, I was interning 1 day a week at UCSF and thoroughly enjoying it, but what was I going to do with the other 4 days? Luckily, my postdoc advisor was extremely supportive and let me continue my postdoc, but I was no longer engaged in the work. I decided to apply for a Scientific Curator position at The Arabidopsis Information Resource (TAIR). The position was perfect as it gave me the opportunity to learn about scientific databases, software development and curation. This turned out to be extremely helpful for managing software inventions in my role at UCSF. Also, TAIR had 1 year of funding available for a plant sciences PhD and were willing to employ me for 4 days a week. After a year at TAIR, a position opened at UCSF and I became a Licensing Associate.

In summary, my career change gave me a new appreciation for the skills I learned as a scientific researcher. Not everyone is suited for an academic career and I feel like I'm making a bigger scientific impact by facilitating the commercialisation of numerous medical technologies than I ever would through managing my own academic lab.

What else can you do if you like disciplined thinking and like to stay informed about science but no longer wish to stay at the bench? Working for a funding council means that you will stay informed about what's going on in science – much more than when you worked on your project. You will gain a broad perspective but lose the chance to contribute yourself. However, you will be able to influence the way money is distributed or ultimately help to shape governmental policies. It can be a very rewarding career as the following story demonstrates. **Dr Steven Hill** is Head of the Research Councils UK (RCUK) Strategy Unit, with the remit to stimulate and encourage collaboration across the seven constituent research councils.

Steven writes:

> As long as I can remember I have always been interested in how the world works, and this was further strengthened by several excellent science teachers at school. At school I was particularly interested in chemistry, especially the practical side of things, but was also interested in the natural world. At university I combined these interests and focused on biochemistry and molecular biology. Being a postdoc allowed me to discover my likes and dislikes about scientific research.
>
> I found two, rather different, aspects of research particularly fulfilling. The first is the problem-solving aspects of designing and interpreting experiments, and integrating the findings of other researchers to draw conclusions. The second was developing research that might have practical applications in the real world, and working with researchers in industry towards those applications.
>
> In terms of negatives, I always found the pace of research very slow – I wanted to know the results of the experiment and did not want to wait! I tend to prefer thinking about the 'big picture' of a problem and sometimes did not enjoy having to focus on the detail, which is essential for really top-quality research.
>
> In retrospect, the most useful skill I gained as a research scientist was the ability to analyse complex and often quantitative information and to extract and simplify the key elements and conclusions. I use this every day in my present role. I would also mention presentation skills, and in particular the ability to present an argument in a convincing way. This is also a skill I use regularly, and it applies as much to one-to-one discussions as it does to presenting to large audiences. Finally, I think writing grant applications and research papers helped to hone my writing skills, especially in terms of clarity and brevity.

With time, I developed an increasing interest and awareness of the importance of the policy environment for ensuring that the benefits of research are realised. This was particularly informed by the issues around the commercialisation of GM crops in the late 1990s. This was my main driver when I decided to leave academia. When an opportunity arose to work as a scientific adviser in the Department for the Environment, Food and Rural Affairs (DEFRA) with special responsibility for GMO scientific advice and policy, it seemed like an excellent opportunity to deploy my skills and knowledge in a Government setting. Transferring to this role was an enormous challenge (more than I expected in advance), but also incredibly rewarding, stimulating and worthwhile.

And so my career path was laid out. I worked as a scientific adviser on GMOs in DEFRA for 2 1/2 years. I then transferred into a different role in DEFRA, leading the secretariat supporting DEFRA's Chief Scientific Adviser. This role involved a much broader range of science (radioactive waste disposal, climate change, animal disease, including major outbreaks of Foot and Mouth disease and Avian Influenza, etc.), and also work on research policy and strategy. After a further 2 1/2 years in this role, I moved to my present job.

I am currently head of the RCUK Strategy Unit. RCUK is the strategic partnership of the seven Research Councils and the Strategy Unit works across the Councils to support and deliver collective working in areas of common interest for the Councils [see mission statement in Box 7.1]. My role involves working with the Councils to develop policy and strategy across the research spectrum. A key aspect is leadership, both leading the Strategy Unit which is a medium-sized team containing 30 people, and also in a wider sense promoting cooperation and collective working across the RCUK family.

All of the skills mentioned above are important in my present role. But more 'people-orientated' skills are also very important. Key skills for me are persuasion and negotiation, and the ability to bring people with you, even if they don't initially agree with your viewpoint. In a relatively senior position, I also need to be able to deal with a wide range of complex issues at the same time, moving focus from one to the other. Part of being able to do this requires a sense of how much of the detail I need to know about a particular issue. You certainly need to be comfortable with not knowing everything about everything!

Box 7.1
Excerpt from RCUK mission statement

The RCUK mission is to optimise the ways that Research Councils work together to deliver their goals, to enhance the overall performance and impact of UK research, training and knowledge transfer and to be recognised by academia, business and government for excellence in research sponsorship.

The overall aim of RCUK is for the UK Research Councils to be recognised as the benchmark around the world in terms of the impact they have and the ways they work.

(http://www.rcuk.ac.uk/aboutrcuk/org/default.htm)

The importance of influencing government policy on science cannot be overestimated. A renowned world expert in nuclear fuels and technology, **Dr Sue Ion DBE FREng** is a member of the Council of Science and Technology and, as such, advises the Prime Minister, and First Ministers of Scotland and Wales, with regard to medium- and long-term strategic policies. She spent 14 years as Chief Technology Officer at British Nuclear Fuels plc (BNFL). Dr Ion, a non-executive board member of the Health and Safety Laboratory, was appointed Dame Commander of the British Empire in the Queen's New Year Honours List in 2010.

Sue writes about her career to date:

> I went to an all-girls grammar school from 1966 to 1973 and was very lucky to have good enthusiastic teachers in the science subjects. I was always interested in Atomic Energy and in fact chose a book on the subject as a prize at O-level which I still have to this day!
>
> I did maths, physics and chemistry at A-level and went to Imperial College to do Materials Science. I stayed on to do a PhD in a topic that was relevant to the nuclear sector. It was a study of the high temperature mechanical properties of the metal alloy clad that encases nuclear fuel elements in the UK's gas-cooled reactors. I liked the idea of researching something new and different; of building on what others had done but making a unique contribution. I was lucky to work in a lab where there were half a dozen PhDs and postdocs at any given point in time and there was great camaraderie; willingness to help each other and endless really challenging debates about our experiments and results. The only thing I was wary of was the uncertainty in career path once I'd finished my PhD.
>
> It was natural therefore for me to seek a career in the Nuclear Industry when I completed my PhD. I wanted to continue in research so joined BNFL's research division in 1979. I spent my early career in research topics and technical support associated with the company's activities in nuclear fuel supply. As a research scientist I gained the knowledge that you should first review what others have done before you embark on a project (it saves a lot of grief and heartache and stops you wasting time and resources!). I used and further developed my ability to problem solve; and gained an ability to appreciate just how valuable technicians and other support staff are.
>
> I led a team of scientists and technologists within 5 years and then a department of some 30 staff. The nuclear sector is very international and I was lucky to travel to countries where the UK had commercial or scientific links. This included Japan, Russia, most of Europe and the US and Canada. From 1987 to 1989 I headed up part of BNFL's Sales and Marketing Division and then spent a year as the Executive Technical Assistant to the company's Chief Executive before returning to a major site as Head of R & D. In 1992 I became Chief Technology Officer for the whole company. This meant I was responsible for over 800 staff on four sites and four laboratories. I also oversaw the building of a new R & D at Sellafield costing over £250m.
>
> During my time with BNFL I also represented the UK on several important international committees. I was a member of the International Atomic Energy Agency Standing Advisory Group on Nuclear Energy and to this day I represent the UK as Chair of the Euratom Science and Technology Committee and as a member of the US

Department of Energy Nuclear Energy Advisory Committee. Since 2004 I have been a member of the UK Council for Science and Technology, I served as a Council Member of EPSRC from 2006 to 2010 and of PPARC from 1994–2001. Since BNFL was wound down in 2006, I have spent time with the Royal Academy of Engineering on science and technology policy matters in the energy sector more generally and I served as Vice President from 2002 to 2008.

I currently hold visiting Professorships at Imperial College and UCL and serve as a Member of the Board of Governors of the University of Manchester. Following the closure of BNFL, I am now an Independent Engineering Consultant and non-Executive Director. Skills that are particularly relevant to my current roles include an ability to understand people and an appreciation of the interplays between government, academia and industry. I derive excitement from finding exceptional talent and from making links between researchers who go on to make discoveries at the interface of their topics.

But how do we find these exceptional talents? Work to help individuals develop and flourish starts early, and needs exceptional teachers. **Dr Scott Crawford** is one of them, teaching at Highgate School, London.

In Scott's own words:

I originally decided to study science because at school I had a super biology teacher who opened my eyes to the possibilities in science. He was rather a maverick in the classroom who thought that pupils should do most of the work on their own. We had to learn a topic independently and then review it with him in class. He pushed us when we needed to be pushed, and supported us when it was appropriate. After a few years in his class, I was certain that I should study biology at university. I went to the University of Oxford and did my undergraduate degree at St Peter's College. With a BA in Biological Sciences under my belt I went to do postgraduate study at the University of York. My supervisor was Professor Ottoline Leyser CBE FRS. There, I discovered my likes and dislikes about scientific research.

I enjoyed working with really intelligent people, who had an infectious enthusiasm for their subject. In the lab, the sense that we were all working towards a common goal was palpable, and personal ambition never came in the way of progress. I relished those moments when an experiment bore fruit, or when a throw-away statement in a paper sparked a new line of enquiry. Travelling around the world for conferences was a real delight; I cannot imagine many jobs where you can journey to New York, Chicago and Marrakesh all in the space of a few years.

Of course, when the experiments did not yield the expected results, or the transformation failed or the PCR reaction crashed mid-cycle, lab-life lost a little of its lustre. My supervisor would always remind me that situations such as those put the 're' into research. In retrospect, the 'failed' experiments often delivered the most interesting results.

During my time in the lab as a research scientist, I developed my project management skills to a very high level. I became tenacious and spirited in the face of adversity, and I learned how to present myself and my work to a range of audiences. I became adept at seeking out collaborations within my own lab and across the department, and after a few years I learned to say 'no' when faced with myriad pulls on my time.

After 4 years in the lab I made a very honest appraisal of my situation. I decided that I did not have the necessary determination to succeed at the highest level in science. I had met many top scientists at my home university and at conferences and I always saw in them an extraordinary determination that I thought I could never emulate or sustain. I therefore thought it better to take the skills and knowledge I had developed to a new arena.

During my time at York, I was involved in various aspects of undergraduate teaching (demonstrating in practical classes and tutoring). I very quickly discovered that I enjoyed teaching more than lab-based work. This notion was further strengthened through my involvement with various outreach activities aimed at enthusing young people with science. With that in mind, I sought employment as a teacher.

During the last year of my PhD, I applied for a job as a biology teacher at a large independent school in London. Within a week of my application I was invited for an interview during which I had to teach a Year 9 lesson. Two days after the interview I was offered the position, which I readily accepted. As the job was at an independent school, I did not have to formally retrain as a teacher (independent schools are exempt from recruiting only qualified teachers).

Currently, I work full-time as a teacher at Highgate School in London (see mission statement in Box 7.2). I teach a broad-spectrum science course to Year 7 pupils, and IGCSE and A-Level biology to pupils further up the school. In all, I teach seven classes amounting to around 20 hours of teaching each week. I am also a sixth-form tutor responsible for 15 Year 12 pupils. I am responsible for the School's Young Enterprise programme, I run the Biology Society and I also help to coordinate the production of the School's termly science newspaper. Outside of the classroom, I am involved in a range of outreach activities to raise the aspirations of children in Haringey.

My day runs from roughly 7.30 through to 7.30. It is a horrible cliché but every day is different. I enjoy the rapid switch from teaching Year 7 pupils something simple in one

---

Box 7.2
About the Highgate School (extracts)

These actions define and determine our ethos:

- Advancing the educational and intellectual and moral development of our pupils, whatever their capabilities, attributes and interests.
- Fostering open-mindedness and thoughtfulness in our pupils and our school

Our aim is to be an academic school and a place for learning and scholarship, where:

- We work to ensure pupils are knowledgeable and are active, independent learners.
- Teaching is rigorous, critical, enagaging and learned.

(http://www.highgateschool.org.uk/)

lesson, such as how to use a Bunsen burner, to explaining something rather more complex in the next, such as action potentials in nerve cells. No matter how prepared you are, there is always an element of the unknown when you enter the classroom. That's what makes it exciting. A teacher never stops learning, and I relish the fact I will never master my job completely. Each lesson is a fresh opportunity to try something new.

Some skills that I developed as a research scientist are particularly relevant to my current job. I learned how to juggle a great number of balls when working towards my doctorate and I certainly juggle every day in school. Planning lessons, marking work, observing colleagues, organising trips, dealing with behavioural issues, meeting parents and, of course, teaching lessons represent only a handful of the balls in the air on an average day.

Having the capacity to approach problems from many angles is certainly a useful trait in a teacher; when a pupil is struggling, one has to explore various avenues to find the approach that works best. Therefore, the problem-solving and creative-thinking processes that I developed as a researcher are skills that are put to use everyday in the classroom.

And there is a wealth of other careers where the skills acquired during your PhD or a spell as a postdoc are valued. The City of London, for example, has always sought to employ scientists and the rewards of a such a career are not just monetary. Here is the story of a City highflyer, **Dr Jessica James,** Managing Director, Global Head of Quantitative Investor Solutions, Citigroup Inc.

I never planned to go into finance. Up to age 12, my passionate desire was to be a vet. After that, I discovered physics, and for the next decade I didn't want anything else. I was at a small convent school, and I was the first girl ever to take any of the sciences at A-level. Fortunately, my physics teacher was rather good, and we spent contented hours talking about the philosophy of quantum mechanics and swapping Tom Sharpe novels.

I went to Manchester University to do my degree, which was hugely enjoyable. Not only was the physics all that I had hoped, the ratio of men to women in the department was about seven to one. After 15 years at my convent school, I was more than ready to discover the opposite sex. My plans in this regard were temporarily stymied by my mother, who booked me into the only convent-run hall of residence in Manchester, but after I worked out how to open my basement window with the fire escape key, there was no stopping me.

I never even considered any path of action other than to do a PhD. The only question was where, and I was overjoyed to gain a place at Oxford to do research into theoretical atomic physics. The ratio of men to women doing PhDs in the Clarendon Lab was even more extreme, but they were generally better looking. Manchester had been in the middle of a 'grunge' period when I was there, which coupled with the grim weather and all the chips, didn't make for healthy complexions.

Oxford presented a sunnier clime, and I have many memories of happy punting trips and picnics in the parks. Also I will never forget my wonderful supervisor, Prof. Patrick Sanders, who in his youth had tutored Stephen Hawking. The professor patiently took me through the mysteries of Parity Non-Conservation, the science of how the Universe is somewhat left handed, and I slowly began to make a contribution to the field.

It became clear to both the Prof. and myself that I was a bit of a contrast to his early, famous student. Newton had a miracle summer, Einstein a miracle year ... I had a miracle fortnight, during which the results which formed the basis of my entire thesis emerged from labyrinthine pieces of code and long brain cudgelling afternoons. I realised as I wrote up my thesis that there were an awful lot of people who were an awful lot cleverer than me. One of them was the professor. Another was one of the experimental physicists who spent patient years designing and running experiments to put the theory of Parity Non-Conservation into practice. He was the cleverest man I had ever met, and he still is. Next year will be our twentieth together.

But what to do next? I loved physics – still do! – but I had firmly decided it was better left to the near-geniuses who held lofty conversations with the Prof. Also, the only way that even these elevated mortals seemed to gain advancement was by living a peregrine lifestyle, swapping continents every 2 or 3 years. I wanted something more settled.

I began to do proactive things like wandering down to the careers centre once a month, as my thesis slowly and painfully gained shape and volume. I was fairly convinced that I was qualified for almost nothing, but hoped that something might arise. The moment that my career turned upon was when the Prof. received a letter from a venerable institution called the First National Bank of Chicago.

That bank doesn't exist any more – it got eaten by a larger predator in one of the many mergers which have taken place over the last few years – but it used to do a lot of complex transactions called derivatives. The name means that the value of the transaction is 'derived' from the value of some underlying security, like a share, or a bond. They were finding that they needed someone with good maths, programming and a can-do attitude to help trade and value the transactions. They had written a letter to 'The Head of the Physics Department, Oxford University, Oxford', which could have landed on one of several desks at that time. To my good fortune, it was directed to my supervisor.

I went to a set of interviews, where I realised that the ability to explain theoretical physics to a bunch of non-mathematical salespeople was more useful than I had realised. I also had to learn to eat sushi in a hurry – it was a novelty at that time and the sales force took me out to lunch to see how I would handle it. Fortunately, I've always been able to eat nearly anything, but it started me off on a very expensive habit.

The next interviews were in Chicago – flown there business class. If I had needed more persuasion to join, that was it. I wish I remembered more about that set of interviews, but the free champagne on the way back was all too much for me.

I began by working on interest rate products. They are mathematically fascinating, and I needed to learn a whole different set of maths – stochastic calculus – that hadn't really come up in quantum mechanics. It was challenging and interesting; who could ask for more? I wrote a textbook on the subject; you can still get it on Amazon, though these days it's going very cheap.

The world turned and the fortunes of the bank rose and fell. It was absorbed by another larger bank, and the interest rate department was cut down. I moved to Foreign Exchange (FX). This is a very different area. English readers will be familiar with jokes that begin 'there was an Englishman, an Irishman, and a Scotsman ...'. Inevitably the poor Irish guy does something humorous and dumb. On the trading floor, the Irishman is always the FX trader.

But this isn't really fair. I know I'm talking my book, as the trading floor would say, but there is a lot more to FX than meets the eye. For one thing, if you work hard, it's possible to get a statistical edge on forecasting the rates. We began a whole new business doing this, selling products based on this ability. I had just been made head of the team, and become pregnant for the second time, when another merger struck.

It was extremely funny being pregnant, and in charge of a team on the trading floor, at the same time. The sales guys and the traders were very nervous that I would suddenly fall to the floor and start making second-stage labour yowling noises. I enjoyed dropping hints about how things could start at any time, but eventually, when the due date approached and maternity policies forced me off the floor, I took myself and my stomach home.

The merger, unfortunately, was no respecter of my delicate situation. I was feeling very content at home with a 1-year-old toddler and a 3-week old baby (you'd think I'd be better at maths!) when the news came. This time the FX department took the hit and I started wondering how to do job interviews with a baby. Could I maybe conceal her about my person? These days, when shoulder bags are large enough to hide medium teenagers in, I might manage it. However, at that time, I had to fess up to having her – I couldn't leave her, as she would not take a bottle. I was her only food source. So, I found out a great deal about the attitudes of the different companies to 'family-friendly' policies. One of them – let's not name names, as they still exist, and could sue me – made me leave her, and my mother, who was looking after her, in the ladies loos, while I did my interview. They didn't offer me a job, and I wouldn't have taken it if they had.

Fortunately, I found a more understanding company; Citigroup, where I still work. The product area of designing statistically based trading strategies suddenly expanded, and once more I found myself in charge of a team doing research and statistical analysis. The job is very different now. I no longer spend hours writing code; I'm better in Outlook than C++. Visiting my teams and customers around the world means that air-miles stack up like the US national debt. And I have learned that building sound systems and processes is just as important as having good ideas. At any one time I will be overseeing several research projects, running a trading book and organising collaborations with universities. Foreign exchange does not sleep – our team covers most of the 24 hours, with its members round the world, and a day will often start early with calls to Tokyo and end late speaking to Sydney.

It's uncertain, fast paced, slightly crazy and above all interesting. I never know what the future holds. But it's been a great ride and I wouldn't swap it for anything.

Pause for thought: What type of career other than an academic one appeals to you at this moment?

How we did it

Whenever possible, we invite alumni from our own departments to come and give presentations to current postdocs. In addition, the University Careers Service organises a day each year when several scientists who have successful careers outside academia give presentations for postdocs. Participants have an opportunity to ask questions and to talk informally to the presenters during lunch and coffee breaks.

Summary

We have presented several stories in this chapter that highlight a small selection of possible careers outside academia. In every case, it is clear that the individuals concerned are still using skills that they learned during their research years, that their enthusiasm and dedication mirrors the vision and strategy of their organisations and that they are enjoying their job.

Selected reading

If you only have time to read one book, make it:
Robbins-Roth, C. (2006). *Alternative Careers in Science: Leaving the Ivory Tower*. London: Academic Press.

8

Applying for a job in academia

It is often said that a candidate is assessed by an interview panel within a minute of entering the room – but you have to get to the room in the first place to even be considered for the job. This chapter looks at the process of applying for a job in academia and provides general guidance on all stages of the procedure – from writing an application through to preparation before the interview. The chapter ends with tips for demonstrating your excellence at interview.

The theory

In what follows we assume that you are looking for your first permanent or tenure-track academic post. In the UK this would be Lecturer, in the US Assistant Professor. There are two ways of getting such a post: either by internal promotion or by applying for an advertised post. As the latter case is more typical, we focus on it here.

As there are not many academic jobs available at any given time, the competition for them is fierce – as many as 100 people can apply for each post. Therefore, you will probably need to write several application letters before you get invited for an interview and you may only get a job offer after several interviews. In order to increase your chances of success you need to search widely, do a lot of preparation and planning and hone your interview skills. Interviewing is a communication process centred on talking and listening. It is different from a conversation as an interview is structured. Traditionally, it is conducted face-to-face, but nowadays people can be interviewed by telephone or via a video-link. While statements of factual knowledge and information on a candidate's attitudes and beliefs can be gathered in all three types of interview, a telephone interview misses non-verbal messages. In this sense both you and the interview panel lose important information.

The practice

The effort you must put into applying for a job can be organised into three stages:

1. Make sure you know exactly what type of job you are applying for.
2. Submit an application form and a well-considered covering letter.
3. Prepare for the interview if invited.

All of the above can occupy a number of weeks.

What job are you applying for?

As advertising is costly, the job advert may be very short but there will be further particulars available. Make sure that you get a copy and read it carefully. What exactly is this position – permanent, fixed-term, full-time, part-time? Have you got the research expertise that is asked for? If not, would your experience count, and would you be able to show that you can adapt to what is asked for? Have you taught before, and if not officially, what type of informal teaching (e.g. supervising research and project students) could you claim to your credit? The job specification will mention 'essential' and 'desirable' traits with respect to both skills and experience (see example in Box 8.1).

Box 8.1

Excerpts from a recent specification for a lecturer position

Essential – a postgraduate degree and good communication skills.
Desirable – an aptitude to teach and experience of teaching in a UK university.
The purpose of this post is to provide teaching in therapeutics within the School of Pharmacy.

Preparing a CV and covering letter

All applications need to be accompanied by a well-considered covering letter that addresses all of the issues raised in the advert and job specification. In particular, you need to make sure that you highlight the most important links between your expertise and the job requirements (see example in Box 8.2). Importantly, however, employers not only look at your current expertise, but at your adaptability and proven aptitude to learn. There is anecdotal evidence that many women think that you have to be 100% on target with regard to the required expertise, whereas most men will apply for a job if their current expertise is only a 60% match to the job specification.

Box 8.2

Example of a covering letter

This extract tells the selection panel why the candidate is qualified for the post in Box 8.1: 'I have diverse teaching experience at undergraduate and postgraduate level. I taught pharmacology to third-year medical students abroad, before coming to Oxford. When I was a postdoctoral researcher in Oxford, I gave lectures and seminars in various fields of pharmacology (autonomic neurotransmission, diuretics, etc.) and took practical classes for first- and second-year physiologists and medical students on various topics. I also supervised undergraduate final-year projects, MSc and PhD students on a daily basis, designing experiments, analysing data and assisting in dissertation/thesis preparation.'
The candidate was successful.

Your CV is your visiting card, and how it looks is one of the factors that will determine whether you are shortlisted or rejected. Make sure that your CV is plainly set out – don't let important things get lost in a sea of trivia. Clearly distinguish different achievements and give a heading to each section so that things can be found at a glance (see example in Box 8.3).

What is essential?

- Contact details: your name, full postal address, email and phone numbers.
- Education: state the title of your PhD thesis followed by information about your tertiary education in reverse chronological order.
- Awards and prizes: if you hold a personal fellowship, put it here, plus things like young researcher of the year, best poster at a conference, students' prize for the best tutor, etc.
- Research experience: if you are employed as a postdoc, give the title of your projects and the name of supervisor(s) as applicable.
- Teaching experience: lectures, lab supervision, informal supervision of project students, etc.
- List of publications: divide into refereed papers, conference papers and other contributions (book chapters, newspaper articles, etc.). Sometimes it can be useful to highlight your four most important papers with a one-sentence description for each.

There are other areas that you may want to highlight: research techniques, your IT skills, languages; conferences organised/attended, contribution to the wider academic community, etc.

Box 8.3
Example of a CV
JANE ALISON LANGDALE

Plant Sciences Dept
University of Oxford
South Parks Rd
Oxford OX1 3RB

Tel: 01865 275000
Fax: 01865 275074
Email: jane.langdale@plants.ox.ac.uk

EDUCATION

PhD (Gene detection using immobilised DNA probes), University of London, 1985
BSc Hons (Upper second class) (applied biology), University of Bath, 1982

AWARDS

Royal Society University Research Fellowship, 1993
SERC Advanced Fellowship, 1990

RESEARCH EXPERIENCE

SERC Advanced Research Fellow, University of Oxford, 1990–1993
Associate Research Scientist, Yale University (T. Nelson, advisor), 1988–1990
Postdoctoral Research Associate, Yale University (T. Nelson, advisor), 1985–1988
Graduate Student, St Mary's Hospital and Charing Cross Hospital Medical Schools,
    University of London (A. D. B. Malcolm, advisor), 1982–1985
Undergraduate Research Student, Cold Spring Harbor Lab., (R. J. Roberts, advisor), 1981
Undergraduate Research Student, Imperial Cancer Research Fund, (B. Griffin, advisor),
    1980
Undergraduate Research Student, Unigate Technical Centre, (G. Stanley, advisor), 1979

TEACHING EXPERIENCE

Lectures and tutorials in plant developmental genetics, University of Oxford, 1990–present
Instructor at Cold Spring Harbor Plant Molecular Biology Course, 1986–1988
Demonstrator in biochemistry and genetics, University of London, 1982–1984

PUBLICATIONS
Research papers
Langdale, J. A., Taylor , W. C. & Nelson, T. (1991). The cell-specific accumulation of
    maize phosphoenolpyruvate carboxylase is correlated with demethylation at a
    specific site 3 kb upstream of the gene. *Molecular and General Genetics*, 225,
    49–55.
Stiefel, V., Ruiz-Avilla, L., Raz, R. *et al.* (1990). Expression of a maize hydroxyproline-
    rich glycoprotein gene in early leaf and root vascular differentiation. *Plant Cell*, 2,
    785–93.

Langdale, J. A., Lane, B., Freeling, M. & Nelson T. (1989). Cell lineage analysis of maize bundle sheath and mesophyll cells. *Developmental Biology*, 133, 128–39.

Langdale, J. A., Zelitch, I., Miller, E. & Nelson, T. (1988). Cell position and light influence C_4 versus C_3 patterns of photosynthetic gene expression in maize. *EMBO Journal*, 7, 3643–51.

Langdale, J. A., Rothermel, B. A & Nelson, T. (1988). Cellular patterns of photosynthetic gene expression in developing maize leaves. *Genes Development*, 2, 106–15.

Langdale, J. A., Metzler, M. C. & Nelson, T. (1987). The *argentia* mutation delays normal development of photosynthetic cell-types in *Zea mays*. *Developmental Biology*, 122, 243–55.

Langdale, J. A. & Malcolm, A. D. B. (1985). A rapid method of gene detection using DNA bound to Sephacryl. *Gene*, 36, 201–10.

Invited reviews

Nelson, T. & Langdale, J. A. (1992). Developmental genetics of C4 photosynthesis. *Annual Reviews in Plant Physiology and Plant Molecular Biology*, 43, 25–47.

Langdale, J. A. & Nelson, T. (1991). Spatial regulation of photosynthetic development in C4 plants. *TIGS*, 7, 191–6.

Nelson, T. & Langdale, J. A. (1989). Patterns of leaf development in C4 plants. *Plant Cell*, 1, 3–13.

Book chapters/symposium proceedings

Nelson, T. & Langdale, J. A. (1993). C4 photosynthetic genes and their expression patterns during leaf development. In *Control of Plant Gene Expression*, D. P. S. Verma, Ed. CRC Press, pp. 259–74.

Nelson, T. & Langdale, J. A. (1989). Differentiation of bundle sheath and mesophyll cells in C4 leaves depends on light and vein placement. In *Current Topics in Plant Biochemistry & Physiology*, 8, 251–60.

INVITED SEMINARS

Invited to talk in several departments in the UK and USA, 1990–present
SEB meeting on Use of Mutants to Study Plant Development, Lancaster, April 1992
AFRC meeting on Plant Molecular Biology, Oxford, September, 1992
Approaches to structure–function relationships in photosynthesis, Dusseldorf, September, 1993

PERSONAL STATEMENT

I did my first degree in Applied Biology, specialising in microbiology. The course was a sandwich course that provided me with the opportunity of working at ICRF and Cold Spring Harbor where I learnt molecular techniques. It became apparent to me that such techniques were powerful tools for addressing biological problems. As a result, I did my PhD in Human Genetics, since technology appeared to be advancing most rapidly in this field. However, I became frustrated by the limitations of working with an organism that cannot be genetically manipulated.

Still eager to address questions that concerned whole organisms, I chose to move into the field of plant development. At the time, Yale University presented the best facilities for such work. I therefore carried out post-doctoral research in Tim Nelson's lab. In addition to Tim's input, this work was influenced by Ian Sussex, who is well known for his work on plant development and by Steve Dellaporta, who is an excellent maize geneticist. I published a number of papers that were internationally recognised as being significant contributions to our understanding of maize leaf development.

After 5 years at Yale I was keen to carry out independent research. I was interested in using molecular and genetic techniques to examine the regulation of maize leaf development. I successfully applied for an SERC Advanced Fellowship that was awarded in October 1990. I took up this post in the Plant Sciences Department at Oxford, where a four-person lab was renovated for my use. Since that time, I have procured external funding to equip the lab and to start my own research group. I also secured a Royal Society University Research Fellowship last year. I have also lectured on plant development to second- and third year undergraduates. I continue to do maize genetics in the USA and have obtained travel and support funding for this work.

My long-term aspirations are to lead an internationally recognised research group and to teach undergraduate and graduate students emerging questions in plant biology. Ultimately, I want a permanent lectureship in plant development.

RESEARCH INTERESTS

The development of higher organisms is characterised by the differentiation of multiple cell types. In plants, cellular differentiation proceeds through the continual interpretation of positional information during growth. How such information is transmitted or received by individual cells is currently an exciting question in plant biology. I am addressing this question through the characterisation of genes that regulate the differentiation of single cell types within the developing maize leaf. The maize leaf is an excellent system for the study of cellular differentiation events, since the final differentiated state is well defined both morphologically and functionally. In addition, maize has a rich genetic history that has provided mutants and mutable systems that rival those found in any other plant system.

The mature maize leaf is composed of sheath and blade regions, which are delimited by an epidermal fringe known as the ligule. As a consequence of cell division patterns, a developmental gradient exists with the oldest cells at the tip of the blade and the youngest at the base of the sheath. Both sheath and blade are characterised by a series of parallel veins running the length of the leaf. Surrounding these veins are concentric circles of two distinct photosynthetic cell types, which compose the majority of non-epidermal tissue in the differentiated leaf. These photosynthetic bundle sheath (BS) and mesophyll (M) cells develop coordinately to interact at maturity in the fixation of CO_2 in the C4 photosynthetic cycle.

Our understanding of events that regulate maize leaf development is limited. However, exciting insights have been obtained through the analysis of mutants that disrupt normal patterns of development. *Knotted* (*Kn1*) and *Rough sheath* (*Rs1*) are heterochronic mutants that convert blade tissue (old) into sheath tissue (young). Genes such as *Kn1* and *Rs1*, that control the timing of developmental programmes, most probably regulate a large number of downstream regulatory and structural genes. Consistent with this hypothesis, the *Kn* gene has been shown to encode the DNA binding homeodomain first identified in *Drosophila* genes that regulate embryo development.

In order to dissect control mechanisms further, we need to identify genes that act later than those that regulate organ development. We would predict that mutations in these 'downstream' regulatory genes would disrupt the development of individual cell types. Using the maize transposable elements *Spm* and *Ac* as insertional mutagens, I have identified one such mutant that specifically disrupts the differentiation of photosynthetic BS cells (unpublished data). Cloning and characterisation of the *bundle sheath defective* (*bsd1*) gene is currently supported by an AFRC grant. The *bsd1* gene is one of only two reported plant genes that are known to direct the differentiation of a single cell type. Clearly, additional genes need to be identified. My near-term goals are thus to:

- Continue an ongoing mutagenesis programme designed to identify mutations that disrupt cellular differentiation in developing maize leaves.
- Carry out molecular and genetic analyses to characterise new mutant phenotypes and to confirm whether the phenotypes result from the insertion of a transposon.
- Clone and characterise the mutated genes.

Our analysis: Clearly, the example above is old; however, the key points to note are:

- All of the most important information is on the first page (in standard format the first page ends after the research papers section of publications).
- The personal statement gives the reader an insight to 'why'.
- The research interests highlight past achievements and outlines future goals.

In addition to your CV, universities often require specific application forms to be filled out. In many cases the application process is online – simple in practice but often associated with pitfalls. Make sure that you prepare the information before-hand in a word document and then cut and paste into the form – that way, if the web session times out on you, you don't have to start again from scratch. Before submitting your application, either online or by email, convert it to a PDF file, so that any formatting cannot be inadvertently changed. For online applications, be aware of any file size restrictions. Keep a copy of your application so that you can re-read it before the interview.

Interview preparation

If you are invited for interview, acknowledge the invitation and confirm that you are available at the time allocated. Normally, a short email or a phone-call is sufficient. Make sure that you read the invitation letter carefully – it may ask you to bring supporting documentation (highest degree certificate, passport for immigration purposes, etc.). It will certainly outline what is expected of you during the interview. Typically, this will involve giving a research talk to the department and giving a short undergraduate lecture on a specific topic. You can prepare a research talk well in advance of any interview (see Chapter 2), but the undergraduate lecture will need to wait until you are informed of the topic. We give generic advice about giving a good undergraduate lecture in Chapter 17.

You can now start doing your homework:

- Know your interview panel – make sure you are familiar with their research areas and, if possible, find out something about individual personalities. Their websites are obvious places to look – even better if your colleagues or your supervisor know them personally.
- Know the department – who is in it? Who will you interact with? What courses are taught? What might you contribute to teaching?
- Know the institution – what are the perceived strengths and weaknesses? Are there any idiosyncrasies (e.g. college system in Oxford and Cambridge)?
- Be ready for the obvious questions and for the less obvious (see examples in Box 8.4).
- Think about the points you want to make about yourself and devise a strategy for ensuring that you make them all. If the opening question at the interview is 'tell us about yourself in 5 minutes', you are in luck – you have been given the opening you were looking for. But what is the panel looking for? Re-read your covering letter and the job advert – you need to link your best qualities to the requirements of the job and support them by an example. If you said 'I am an enthusiastic teacher' in your letter, at interview you want to demonstrate this – 'I got so involved in teaching part of a second year thermodynamics course that I have completely re-written it in my spare time.' If you said that you are a hard working person, in the interview you could say something like 'This project required so much attention that often I was the last person to leave the lab.'
- Prepare a question or two to ask the panel at the end of the interview. You could ask the following: is there a start-up grant from the department and if so how much would that be? What conditions have to be met during probation/for tenure? Is there an opportunity for doing multidisciplinary research in the university? Are there any existing networks that I could join? What courses would I

Box 8.4

Example of academic interview questions

- Why do you want this job?
- Where do you want to be in 5 years time?
- What are your most significant achievements to date?
- What do you think you can bring to the Dept (in terms of expertise, money, research connections)?
- What are your strengths and weaknesses?
- What will you do if you don't get this job?
- How well do you perform in a team?
- What is your teaching experience?
- What support would you expect from the department?
- What were your best and worst projects and why?
- What will you do when you've completed your 5-year research plan?
- How will you implement your ideas?
- How will you build your team?
- How big do you want your group to be?
- What are the risks inherent in your project?
- What is your timescale for setting up the group?
- Do you want to teach and, if so, what?

---

be expected to teach in my first year, and in the following years? Will I be asked to supervise undergraduate laboratory practicals in my first year?

- Rehearse your research talk and the undergraduate lecture, even if only in front of a mirror. To overcome potential conflicts between your file and/or computer and the data projector, make back-up PDF files of your presentations.

- Think about how you are going to get to the interview. You want to arrive as relaxed as possible and well ahead of time. Depending on the distance between your place and the place of the interview, you may need to leave very early or a day ahead. Are you going to use a car or public transport? Or walk? Find directions on the internet and look at time estimates. If you go by car, will there be a parking space reserved for you? If you go by train, do you need to change trains and what would happen if you miss a connection?

- Think about what you are going to wear. Of course, what you say is the most important factor at interview, but first impressions count. Research shows that visual clues contribute up to 70% of how a candidate is perceived by an interview panel. Hence you want to be as comfortable as possible and yet look professional. There are no strict sartorial rules, but you should look smart.

Interview tips

An interview can be formal or informal. The interview itself can be short – 45 minutes to an hour – or last several hours or a couple of days. If you are invited for a whole day and are being shown around by members of the department after the official interview, consider this as a continuation of your interview. People will be asked afterwards what they thought of you.

- Typically there will be four to six shortlisted candidates. If you are the first to be interviewed, the panel will be just warming up – they may not have worked together before and so will be working out their own dynamics. This may give you an opportunity to take more of a lead in the discussion. In contrast, if you are the last to be interviewed, the panel may be getting tired and/or restless. In this case you will have to be more dynamic than those who went before you, to maintain equivalent interest.

- When you enter the room, make eye contact with each panel member and if possible shake their hand. They will undoubtedly be introduced to you – so try to put names to faces – of course, you will have researched the names before. During the interview don't fidget even though you may feel nervous; smile occasionally, but do not have a fixed smile on your face; watch your gestures – too much hand-waving gives an impression of uncertainty. Make eye contact with the interviewers – don't stare at your lap. And talk to the person who has asked you a question rather than try to engage the whole panel.

- If you try and avoid a tricky question by diverting to a tangential topic, ensure that you make a very clear point. If you don't, it will be seen as a clear evasion.

- Be prepared to bounce questions back to the panel, either to stall for time or to steer the discussion to your agenda. Do not be afraid to ask a panel member either to repeat a question or to phrase it differently. This is important – especially if a question is long and seems to contain several questions in one.

- It is wise to admit when you don't know something, but do not get defensive.

- Make sure you know how and when you will be told the result of the interview if this information is not offered, and thank the interviewers for the invitation.

- After the interview, make time to reflect on the process so that you learn from it whatever the outcome (see example in Box 8.5).

Box 8.5
The interview experience

'The interview was the most daunting aspect of the whole process. To commence, I gave a 5-minute summary about myself and my work to date – an opportunity to let the six panel members know why I thought I would excel in their department. After my introduction, each panel member had his or her opportunity to ask more direct questions. For example, the Head of Teaching questioned my lecturing experience, while the Head of School probed my motivation for moving to the University. In contrast, the Head of Research was more interested in my Nobel Prize aspirations! Throughout the interview I tried to remember to sit upright and maintain eye contact. In my mind I believe I listened as carefully as I could to the questions asked and took a few seconds to think before replying with a thoughtful answer. In practice, however, I wonder if I was as composed as I recall!

I have yet to hear whether my interview was a success. If not, then I will definitely feel a sense of disappointment; the interview is an intense, soul-baring process. However, no matter what the outcome, I will also see it as a useful learning experience.'

On this occasion, the candidate was unsuccessful but is currently applying for other positions.

How we did it

We have run this workshop in two ways. In the first, we had an introductory talk about different parts of the application process, illustrated by examples. Participants were asked to bring their CV and to give each other feedback in pairs, taking into account points about the 'model' CV. Then we had mock interview practice: we asked participants to form groups of three, label themselves A, B and C and look at the interview questions. They were given 5 minutes to prepare their answers. Then it went as follows: A interviewed B, C observed – 5 minutes; B interviewed C, A observed – 5 minutes; C interviewed A, B observed – 5 minutes. After each interview, the participants from different groups were then asked to share feedback with each other – also for 5 minutes. The seminar leader had three sets of interview questions that were revealed after each interview – so there was no advantage in learning how to answer a given question.

In the second, a panel of academic staff members conducted interviews, having received applications form the participants in response to a 'mock' advert. All members of the workshop observed the panel's discussion before the interviews, each other's interviews and the panel's summary discussion and decision-making process.

Summary

The secret of a successful job application lies in thorough preparation – find out as much information as possible about your prospective employer, write a well-considered covering letter, and put together a clear and concise CV. At interview, try to relax and be yourself – be professional, be honest and enjoy the opportunity to visit another university and to meet interesting people. Regardless of the outcome, you will learn a lot – both about the application process and about yourself.

Selected reading

There is no particular book that we recommend. Instead, you are advised to make contact with the Careers Service in your university or place of work and to use their resources. Try to arrange a mock interview before going for a real one. For actual job adverts, search the general science job websites and the websites of institutions you would like to work for.

9

Applying for an independent research fellowship

In the last two decades there has been a change in the typical career path in academia, particularly in the UK. Previously, the normal trajectory would have been from PhD through one or two postdocs, to probationary lecturer/tenure-track Assistant Professor, and then to a tenured post. Nowadays, however, a favoured intermediate step between a postdoc and an academic job is an independent research fellowship (IRF). In this chapter we offer guidance on applying for IRFs and provide details of the various fellowships currently on offer.

The theory

An IRF allows you to build up an independent research programme before taking on teaching and administration commitments. The best fellowships range from 5 to 8 years – enough time to establish your own research group, break into a new field or map out a new one, and lay solid foundations for your future research programme. Importantly, applying for academic jobs from this position allows prospective employers to make a concrete assessment of your ability to gain research funding, to manage a research group and to publish independent research papers. Assuming that you have proven ability in these areas, you are in a strong position when applying for academic jobs – not only do you already have a good track record, but your employer knows that you will have the time and professional maturity to take on new teaching and administration duties. If you do not have proven ability in these areas, the fellowship has served to allow you to realise this for yourself.

The practice

Most of the advice given in Chapter 8 in relation to submitting an application and attending an interview for an academic job is equally applicable here. Similarly, some of the guidance on grant writing in Chapter 5 applies equally to writing a fellowship application. However, there are also specific issues to consider. For example, before applying for a fellowship you need to identify an institution that will host you if you are successful. The choice of institution needs to be carefully considered because the fit between your research and your proposed research environment is a major criterion upon which applications are judged. You need to think about both the facilities available in a particular department and the potential for academic interactions and collaborations. Having identified a possible destination you will need to contact the Head of Department and ask them if they are willing to support your application. It is likely that they will want you to visit the department and give a seminar and/or talk to the academic staff. This will allow you, and them, to determine whether it would be a good place for you to work.

> **Pause for thought:** Where, other than in your existing department, could you carry out your current research programme? What would be the advantages/disadvantages of working elsewhere?

Having identified a host institution, you can start thinking about your writing your proposal. The key thing to remember here is that it is an *independent* proposal. It must therefore be easily distinguished from your current supervisor's research programme – and not just in appearance. Make sure that you have discussed your intentions with your supervisor – do you agree on which parts of the project you will each take forward? Can either of you see potential areas of conflict? It is important to resolve any differences in opinion before you submit an application – whilst you do not need your PIs support, you will certainly do better if you have it.

> **Pause for thought:** In what direction could you take your current project that would allow you to carve an independent niche?

So, what are the ingredients of a successful application for a personal fellowship? We asked a senior academic, who gets to see a lot of IRF applications, for his views. Tim Softley is Chair of Chemistry and until recently was the Associate Head of Division (Academic) in the Mathematical, Physical and Life Sciences (MPLS) Division of the University of Oxford. He took the lead in educational policy and strategy development for the Division, having oversight of 3300 science undergraduates and 1600 postgraduates. Tim's advice on fellowship applications is as follows:

Independent research fellowships are intensively competitive in general, especially Research Council and Royal Society Fellowships. In my view you need to:

- *Show that you have a vision* – how does what you want to do fit with the 'big picture'? Is this science that, if achieved, could be described as internationally competitive or internationally leading; as having a major or significant impact on the field? If you are not moving from your current department, will the fellowship nevertheless offer you a new opportunity, more independence and a new direction? Will the new independence gained enable you to develop an international reputation in your own right?
- *Provide convincing evidence for the viability of your plans* – make sure that any feasibility studies/calculations are done before you submit. Be realistic about what can be achieved, and make sure the project is properly costed. Have a clear plan of action with reasonable and realistic timescales and targets. Show that there is flexibility in the proposal if the immediate aims do not go according to plan. Explain why the proposed host institution is the best place for what you want to do (as opposed to the most convenient place). Plan the training of the researchers whom you will supervise.
- *Get your track record in good shape* – first author publications in good quality journals are important. 'Published' is much better than 'in preparation' or 'submitted'. Publishing in a good spread of well-respected journals including (but not exclusively) high-impact journals will help build your case.
- *Get your support lined up* – choose your referees carefully (if referees are required) – their *strong* and *unreserved* support is crucial. Ask them whether they are willing to support you strongly, not just whether they will write you a reference. Or, ask them whether they think you should apply for this fellowship. Identify and speak to your potential collaborators and make sure that they are ready to support your application if needed. Make it clear in your proposal what they will add to the project.
- *Prepare proposals and presentations very carefully* – you need to allow considerable time to prepare your application, to proof read it carefully, to seek other people's advice, to practise any presentation you have to make and to have a mock interview with someone who has looked at your proposal. You need to be able to show that you are really excited both about what you have achieved already and what you plan to do. A high level of ambition is good, but be realistic. Make sure your proposal, at least in its introduction and motivation, is accessible to people outside your immediate field.

How we did it

We considered IRF applications in the same workshop as grant applications (see Chapter 5 for details). We also carried out mock interviews for course participants if they requested them.

Selected reading

For detailed information about fellowships and guidelines, please see the respective websites:

1851 Research Fellowships provide three years funding to young scientists or engineers of exceptional promise to conduct a research project of their own instigation – http://www.royalcommission1851.org.uk/res_fellow.html

BBSRC David Phillips Fellowships provide 5 years' funding for postdocs who have demonstrated high potential and who wish to establish themselves as independent researchers – http://www.bbsrc.ac.uk/funding/fellowships/david-phillips.aspx

Daphne Jackson Memorial Fellowships offer two years' funding to allow talented scientists and engineers to return to careers after a break of 2 years or more – http://www. daphnejackson.org/

EPSRC Career Acceleration Fellowships provide up to 5 years' support for outstanding researchers at an early stage of their career – http://www.epsrc.ac.uk/funding/fellows/ careeracc/

EPSRC Postdoctoral Fellowships enable talented new researchers to establish an independent research career shortly or immediately after completing a PhD. They provide funding for up to 3 years – http://www.epsrc.ac.uk/funding/fellows/Pages/postdoctoral.aspx

Leverhulme Early Career Fellowships aim to provide career development opportunities for those who are at a relatively early stage of their academic careers, but with a proven record of research. Awards contribute 50% of the fellow's salary and it is expected that the Fellowship will lead to a more permanent academic position – http://www. leverhulme.ac.uk/funding/ECF/ECF.cfm

NERC Postdoctoral Fellowships provide up to 3 years' funding to support outstanding environmental scientists as they become independent investigators – http://www.nerc. ac.uk/funding/available/fellowships/typesofaward.asp

Royal Academy Engineering Fellowships are designed to promote excellence in engineering. The scheme provides funding for 5 years to encourage the best researchers to remain in the academic engineering sector – http://www.raeng.org.uk/research/researcher/postdoc/ default.htm

Royal Society Dorothy Hodgkin Fellowships support excellent early-career scientists, who require a flexible working pattern, for up to 4 years – http://royalsociety.org/Dorothy-Hodgkin-Fellowships/

Royal Society JSPS Postdoctoral Fellowship Programme provides opportunities for early career researchers from the UK to conduct research in Japanese institutions – http:// royalsociety.org/JSPS-Postdoctoral-Fellowship-Program/

Royal Society Newton International Fellowships attract the world's best early-career researchers to the UK for 2 years – http://royalsociety.org/Newton-International-Fellowships/

Royal Society University Research Fellowships provide outstanding scientists, who have the potential to become leaders in their chosen field, with the opportunity to build an independent research career. Funding is available for up to 8 years – http://royalsociety. org/University-Research-Fellowships/

Part II

Thriving in your new job

Your living is determined not so much by what life brings to you as by the attitude you bring to life; not so much by what happens to you as by the way your mind looks at what happens.

Kahlil Gibran

The first part of this book concentrated on establishing yourself as an independent scientist who can confidently communicate your research, to the extent that you can secure a position as a principal investigator (PI). In this section we focus on working with others and on building your research group. In the first chapter (Chapter 10) we look at ways to understand your personality profile – this knowledge will help you learn how best to interact with others. You can also learn from those who have gone before you, and in Chapter 11 we hear from some young PIs about their experiences of the postdoc–PI transition. We then turn to the thorny issue of managing people (Chapter 12), and on to recruiting and supervising PhD students (Chapter 13) and postdocs (Chapter 14). Partly because scientists have no formal management training, but also because research is very difficult to carry out unless there is a degree of harmony in the research team, the ability to manage people can make or break a young PI's career. Similarly, the way in which you interact with others through networking and collaboration, and the extent to which you contribute to the scientific community can influence how you and your research is perceived. We discuss how to make the most of networking and collaborations in Chapter 15. Teaching and learning figure prominently in the final four chapters both in formal classroom settings (Chapters 16,17 and 18) and in the mentoring process (Chapter 19).

10

Handling new roles

You have a new job, possibly a new home and certainly new responsibilities. This is an exciting, challenging and often daunting phase of an academic career. How can you ease the transition? This is a good time to take stock of your strengths and weaknesses. What are you naturally good (or bad) at? Are what you perceive as your strengths seen as strengths or weaknesses by those who work with you? Knowing the answers to these questions can help you plan the early stages of this career phase so that you maximise your potential. In this chapter we discuss the general concept of Belbin profiles and what each classification means. More importantly, we discuss which roles within academia will come naturally to each profile and which will need more effort.

The theory

The previous chapters have emphasised the development from dependence on others to independence: how to increasingly take a lead in actions and have more responsibilities. Thus far, interacting with others has been in the background, yet scientists increasingly work in groups and sometimes in large multinational teams. To be effective in the context of a team requires an understanding of what you are naturally good at and an acceptance of the fact that some things are better left to others.

In the last century there was considerable research into effective team-working strategies. In the 1980s and 1990s Meredith Belbin, from the Henley Management School, proposed a model based on 'team roles' (Belbin, 1981). In his words:

The term 'team role' refers to a tendency to behave, contribute and interrelate with others at work in certain distinctive ways. For practical purposes, one needs to discriminate sharply between a person's team role and 'functional role', where the latter refers to the job demands that a person has been engaged to meet by supplying the requisite technical skills and operational knowledge.

This distinction between team and functional roles is helpful in the case of a conflict between a person's natural inclinations and job requirements; it can help in renegotiating the role. While originally developed in the context of management teams, this model was subsequently shown to be more generic and valid for non-management teams (Belbin, 2003). A recent review summarises progress since the inception of the original hypothesis (Aritzeta *et al.*, 2007).

According to Belbin there are nine roles in a work situation. In the context of Chapter 1 the role would be 'work colleague' and Belbin's categories would help to define sub-roles. These are known as completer finisher (CF), implementer (IMP), team worker (TW), specialist (SP), monitor evaluator (ME), coordinator (CO), plant (PL), shaper (SH) and resource investigator (RI) (see Box 10.1). Each role is

---

Box 10.1
Team role descriptors, strengths and allowed weaknesses

Team role	Descriptors	Strengths	Allowed weaknesses
Completer finisher (CF)	Anxious, conscientious, introvert, self-controlled, self-disciplined, submissive and worrisome.	Painstaking, conscientious, searches out errors and omissions, delivers on time.	Inclined to worry unduly. Reluctant to delegate.
Implementer (IMP)	Conservative, controlled, disciplined, efficient, inflexible, methodical, sincere, stable and systematic.	Disciplined, reliable, conservative and efficient, turns ideas into practical actions.	Somewhat inflexible. Slow to respond to new possibilities.
Team worker (TW)	Extrovert, likeable, loyal, stable, submissive, supportive, unassertive and uncompetitive.	Cooperative, mild, perceptive and diplomatic, listens, builds, averts friction, calms the waters.	Indecisive in crunch situations.
Specialist (SP)	Expert, defendant, not interested in others, serious, self-disciplined, efficient.	Single-minded, self-starting, dedicated; provides knowledge and skills in rare supply.	Contributes on a narrow front only.

Team role	Descriptors	Strengths	Allowed weaknesses
Monitor evaluator (ME)	Dependable, fair-minded, introvert, low drive, open to change, serious, stable and unambitious.	Sober, strategic and discerning, sees all options, judges accurately.	Lacks drive and ability to inspire others.
Coordinator (CO)	Dominant, trusting, extrovert, mature, positive, self-controlled, self-disciplined and stable.	Mature, confident, a good chairperson, clarifies goals, promotes decision making, delegates well.	Can be seen as manipulative. Offloads personal work.
Plant (PL)	Dominant, imaginative, introvert, original, radical-minded, trustful and uninhibited.	Creative, unorthodox, solves difficult problems.	Too preoccupied to communicate effectively.
Shaper (SH)	Abrasive, anxious, arrogant, competitive, dominant, edgy, emotional, extrovert, impatient, outgoing and self-confident.	Challenging, dynamic, thrives on pressure, has drive and courage to overcome obstacles.	Prone to provocation. Offends people's feelings.
Resource investigator (RI)	Diplomatic, dominant, enthusiastic, extrovert, flexible, inquisitive, optimistic, persuasive, positive, relaxed, social and stable.	Extrovert, communicative, explores opportunities, develops contacts.	Over-optimistic. Loses interest after initial enthusiasm.

Source: Aritzeta *et al.* (2007) after Belbin 'Team roles at work' (p. 22).

characterised by desirable strengths and allowable weaknesses. Almost everybody has a strong preference for one or two primary roles, but these roles may not necessarily be their most effective ones. In reality, there is likely to be a mix of three or four roles that you have to play in any group or a team.

The different roles are grouped as follows: Action orientated – SH, IMP, CF; People oriented – CO, TW, RI; Cerebral – ME, PL, SP.

Pause for thought: Have a look at descriptors of Belbin roles in Box 10.1. Can you define any common ground for roles in each group?

The practice

Belbin's approach has been very successful, and many institutions now ask employees to obtain a 'Belbin profile' to aid project management. In some cases, prospective employees are required to submit a profile before being interviewed for a particular job. The online version (http://www.belbin.com/) is straightforward to use, and has two options: self-perception and/or assessment by up to five other 'observers'. For the self-assessment, users fill in answers to a series of questions that are divided into three categories: (1) What I believe I can contribute to a team. (2) If I have a possible shortcoming in the context of teamwork, it could be that ... (3) When involved in a project with other people, I ... Observers have a different format whereby they describe a colleague's behaviour with words from two lists of adjectives (positive/negative). In both cases a report is generated and then emailed back.

The full Belbin test gives insight into your behaviour, strengths and weaknesses in work situations. How can you benefit from the information generated? One of the most important things to check is whether your self-perception equates to how others see you. In Box 10.2 there is an extract from a recent online test (the anonymity of the participant is preserved).

Pause for thought: Look up the description of roles in Box 10.1. Without doing the Belbin test, write down your two most and least preferred team roles.

Two other reports to analyse are the counselling report (Box 10.3) and character profile (Box 10.4). These are shown for the same individual as in Box 10.2.

Pause for thought: Think about your career to date. In which work situations have you thrived, and in which you were miserable?

Academia and Belbin roles

How could such a test be useful for you? It is not customary in academia for search committees to look for suitable team members – instead, the most common mantra is 'pick the best scientist and let them do what they want'. We think it is useful because the Belbin profile tells you a lot about how you relate to others in a work situation. However, you need to keep in mind the following general principles:

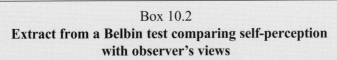

Box 10.2

**Extract from a Belbin test comparing self-perception
with observer's views**

The most striking difference in the example shown in Fig. 1 is the perception of primary roles. Self-perception leads to almost equal weight between Resource Investigator (RI) and Plant (PL) whereas the observers do not recognize the RI role in the same way. Other discrepancies are evident for roles of team worker (TW), Implementer (IMP) and completer finisher (CF), whereas for other roles there is broad agreement.

PL (Plant)
RI (Resource Investigator)
CO (Coordinator)
SH (Shaper)
ME (Monitor Evaluator)
TW (Team Worker)
IMP (Implementer)
CF (Completer Finisher)
SP (Specialist)

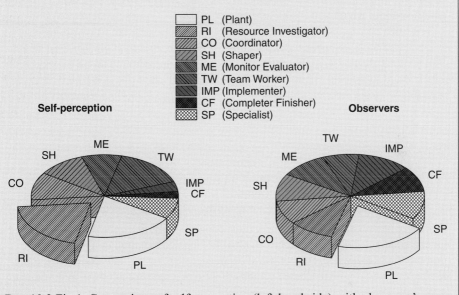

Box 10.2 Fig 1. Comparison of self-perception (left-hand side) with observers' views.

- The profiling places emphasis on teamwork. Thus, it assesses the contributions you make to a team.
- Any organisation (be it a lab group, a department or a university/company) needs a balance of individuals with each team role profile. Thus, one type is no better than another type – they are just different.
- Your individual profile can change over time in response to different environments.
- Your profile should be used to assess where you and other team members are coming from when tackling different issues. Being aware of your own approach and that of others means that you will be able to get the best out of situations.
- Pay special attention to mismatches between your self-perception and observer rankings. You make your mark in the areas most appreciated by other people.

Box 10.3
Extract from a counselling report

Yours is essentially a pioneering profile. You are one of the few people equally ready to develop new ideas on your own or in conjunction with others. Your best line of work is one in which you are required to explore possibilities and to take advantage of new opportunities. You have some features of the visionary. But take care you do not become isolated from others and resistant to the contributions they can make to the development of what is new. The implications for career development are that you need to steer yourself towards areas where change is highly valued. That would provide an environment in which you could flourish.

Your own perception of your top team role is supported by the views of others.

On a final note, you need to take account of the role for which you are least suited. You do not appear to have the characteristics of someone who attends to the details on which every significant operation can hinge. So if you can work in harmony with someone who has these qualities, your own performance is likely to benefit.

Box 10.4
Extract from a character profile

Strengths

Generally regarded as an individual who is caring, clever, disciplined, encouraging of others, good at follow-through, knowledgeable, loyal, observant, outgoing and professionally dedicated. Has innovative tendencies and needs to work in a mentally challenging environment. Could be good at solving complex problems or at introducing new ideas. Performs best on her own or when in a team is given freedom to exercise her particular skill.

Requires work where she can use her outgoing nature. Likes to use personal initiative. Suited to work where she can use her acquired knowledge. Keen to follow a professional career path. Keen to explore and develop new ideas. Needs to work in an environment that offers scope for personal expression.

Possible weaknesses

Dependent on continuous stimulation and inclined to lose interest quickly.

How do Belbin profiles relate to situations typically encountered in academia?

- As a postdoctoral researcher working in a group, you are expected to carry out research, to be very good at that research and to co-exist harmoniously with others in the group. If you are on your way to independent research, you are most likely to be rewarded for the following attributes: plant (PL), specialist (SP), monitor evaluator (ME), and completer finisher (CF). That is, coming up with ideas, having specialist skills, good judgement and able to see a job through to the end (publishing). If you are a team worker (TW), that is an extra bonus. In this environment, a shaper (SH) can be extremely constructive or extremely destructive depending on group dynamics and on the principal investigator (PI). If the PI is a good manager, a SH will be beneficial to the group as a whole. Those with resource investigator (RI), coordinator (CO) and implementer (IMP) attributes are most likely to be 'exploited' by PIs because they are the ones who can hold the group together in the absence of regular contact with the PI, particularly if they also have SP skills. However, they are probably not promoting themselves enough to ensure a successful transition to the next career stage and may unwittingly find themselves stuck on the career postdoc track.

- As you establish your own research group, PL and SP attributes continue to be rewarded as you apply for research grants. However, CF and IMP attributes become more important as it is likely that you have to be more self-disciplined than previously. When you start to recruit individuals, you are at an advantage if you have CO attributes because you will get the best out of everyone and thus your career will benefit. You should also realise that your team will work best if there is a balanced profile of roles (something to think about when recruiting).

- As a PI, you will also be part of a larger team (e.g. in a department) and you need to think about your role in that context. Are you the TW who shows up to help out on open days? Are you the SH who designs a new MSc course? Are you the CF who goes through the end-of-year accounts? Are you the SP who runs the microscopy unit? Are you the CO who organises the annual party? Are you the IMP who oversees the plant growth facilities?

- You have a further role to play in your research field. Are you the RI that initiates a collaborative project or the CO that makes it happen? Are you the SH who influences the international research agenda? Are you the SP who provides an invaluable community resource?

Bottom line – at some stage you may have to perform all nine roles. In stressful situations, however, people tend to revert to type. Therefore, make sure you are tackling the projects that are suited to your type in order to maximise the impact you make.

Myers–Briggs personality profiling

While Belbin profiles are well known, they are not the only personality profiles that can be used for self-assessment and for assessing your suitability for a given team role. Another type of test is based on the theories of Carl Gustav Jung, as developed further by Briggs and Myers. This test is administered as the Myers–Briggs Type Indicator (MBTI) and is available on the internet at various websites, e.g. http://www.teamtechnology.co.uk/mmdi.

The psychological assumption underlying this test is that people are either born with, or develop, certain ways of thinking and acting. There are four dichotomies: extraversion (E) ↔ introversion (I), sensing (S) ↔ intuition (N), thinking (T) ↔ feeling (F) and judgement (J) ↔ perception (P). This theory is non-judgemental, these are simply differences; the working hypothesis is that individuals may prefer one combination of these differences. This leads on to 16 different personality types that are recorded by a combination of four letters, for example ISTJ.

How reliable are these tests? We looked at a Mental Muscle Diagram Indicator (MMDI™) test that is based on self-perception only. The report is meant for self-understanding in a team setting. It warns explicitly against using the results for selection or recruitment purposes, as the questionnaire does not assess your competence or preference for certain types of jobs – they may be divergent.

We asked the same person who did the Belbin test above to do the MMDI™ test. The results are shown in Box 10.5.

Box 10.5
Results of a Myers–Briggs test

- Pattern matching – ENFP (or INFP) as two 'whole' types were close within 3% difference, with slight preference for ENFP (80%).
- An independent scoring for each personality type. In this method the results for ENFP were: *E*xtroversion 52%, i*N*tuition 58%, *F*eeling 62% and *P*erception 60%. Elements of the INFP profile were as follows: *I*ntroversion 48%, *S*ensing 42%, *T*hinking 38% and *J*udgement 40%.
- Preferred team roles (in descending order): exploring, campaigning, harmonising, activating, innovating, analysing, conducting and clarifying.

The extravert type ENFP is described in the MMDI™ report as: 'ENFPs sense the hidden potential in people. They enjoy starting discussion of activities that challenge and stimulate others into having new insights about themselves. They are enthusiastic about new projects or causes that offer the potential for a beneficial impact on people, especially when it involves breaking new ground.'

In contrast, INFPs 'have a strong sense of the hidden principles that govern how the world works. They are interested in theoretical models and explanations, and when other people put forward their own theories they put them to the test to find out how true or robust they are. They enjoy solving difficult intellectual problems and seek to understand the real truth behind any situation, even when it involves several complex factors.'

Our tester thought that both profiles reflected her personality, and that in different situations she would 'use' one or the other. We also thought that there was a broad agreement with the results of the Belbin test (Boxes 10.2–10.4).

How we did it

Two weeks before the workshop, we sent a link to participants and asked them to fill in the online Belbin profile (http://www.belbin.com/). We recommended doing the full profile and thus having at least four work colleagues or friends who were willing to act as assessors. The test results were emailed to one of us, test administrators, and then forwarded to participants. For confidentiality reasons, we only discussed individual Belbin profiles one-on-one. In the workshop, we covered general principles and the theme of academia and Belbin. Some participants were willing to talk about their expectations and the test outcome; however, these contributions were not solicited by us.

Summary

In this chapter we have introduced and illustrated Belbin team profiles. Knowing your Belbin profile helps you to assess realistically your strengths and weaknesses so that you can maximise your performance at work. This, in turn, should ease the transition from being a postdoc to a PI.

Selected reading

If you only have time to read one book, make it:
Belbin, R. M. (1981). *Management Teams: Why They Succeed or Fail*. London: Heinemann.

Other texts:
Aritzeta, A., Swailes, S. & Senior, B. (2007). Belbin's team role model: development, validity and applications for team building. *J. Managem. Studies*, 44, 96–118.
Belbin, R. M. (2003). *Team Roles at Work*. London: Butterworth-Heinemann.

11

Learning from other people

You will have just made the transition from being one of the most academically able and experienced postdocs in a research group to being the most inexperienced PI in the department. Although you only really learn from your own errors, you can avoid mistakes if you know ahead of time what they are likely to be. In this chapter we asked four young PIs to recall their experiences of the postdoc–PI transition and to offer their advice for avoiding problems – the advice is remarkably consistent.

The theory

There is a constant tension between change and routine, and in the context of an academic career it is perhaps most visible during the transition from postdoc to PI. Before we turn to the practicalities of doing well during this period, it is worthwhile considering the wider background. Life *is* change – stars are born, live and die on such a long timescale that humans cannot even imagine, and at the other end of the scale processes in living cells occur over nanoseconds. However, as the popular saying goes, 'only busy cashiers and wet babies like change'. Change unsettles us and our biological reaction is of the 'fight or flight' type. But change also brings renewal and opportunities, and in a professional environment, if it is tackled well, it can strengthen your career prospects for years to come.

As seen in Chapters 8 and 9, you can start your independent research career either in the context of an academic job or as an independent research fellow. Although the two positions carry different duties and expectations, they share one thing in common – time has to be managed *differently* from when you were a postdoc. While many postdocs work very hard, they often work at hours to suit themselves (certainly not 9–5). However, as a PI you have to fit in with other people – either in the context of time-tabled lectures and meetings or to oversee the work of your research group.

The practice

Independent research fellows

At face value, the transition from postdoc to IRF should be straightforward and *fun* – after all, you have just been given funding to carry out your *own* research for the next few years (often for a time period that is longer than for any position you have held previously) and you have enough money to pay other people to help you. You have no-one telling you what to do, no teaching and no administrative duties – what more could you want? To gain insight into this transition, we asked two IRFs the following questions:

1. How long have you been a PI?
2. What did you do before?
3. Have you got any grants and/or people working with you?
4. What did you find most difficult in the transition from postdoc to independent researcher?
5. What advice would you give people applying for fellowships/lectureships?
6. What advice would you give to new PIs?

Jamie Warner is a Royal Society University Research Fellow in the Department of Materials Science at the University of Oxford. His research interests include the design, synthesis, characterisation and understanding of novel nanostructured materials. He is particularly interested in the unique quantum mechanical properties that arise in materials when they are made very small. This involves carbon-based nanomaterials such as fullerenes, metallofullerenes, single-walled carbon nanotubes, multi-walled carbon nanotubes, peapods and graphene.

Jamie writes:

> I started my first IRF in October 2008. This was a 3-year fellowship called the Violette and Samuel Glasstone Fellowship in Science. I held this for 2 years and then started a Royal Society University Research Fellowship in October 2010. Before this, I was a postdoc in the Department of Materials for 2 years, under the supervision of Professor Andrew Briggs. Before that I was a postdoc at Queensland University of Technology in Australia for 8 months, and before that a postdoc at Victoria University of Wellington in New Zealand. I did my undergraduate and PhD in the Department of Physics at the University of Queensland in Australia.
>
> My funding for running the lab comes mainly from the Royal Society but I am also a co-investigator on an EPSRC-funded project called 'Putting spin into carbon nanomaterials'. I currently supervise three PhD students and one fourth-year undergraduate research student. I have one postdoc who I directly manage, and two others whom I line manage.
>
> For me, the most difficult part of the transition from postdoc to PI was creating an identity within the department that distinguishes me from others. Another challenge

was setting up labs from scratch. Developing management skills was also not easy and neither was the transition from being a researcher in the lab to being a manager with lots of meetings and little time for lab work.

My advice to people who are thinking about applying for fellowships is to build a strong portfolio of publications in the area that you want to apply for. In writing your applications be visionary and outward looking, set up international and national colla-borations, link theoretical work with experimental work, be imaginative and creative and don't rehash other people's ideas – have your own and be daring.

Once you start as a new PI, make sure to get advice from several mentors and be in close communication with your head of department. As your research (and group) expands, there are a number of things I would suggest. Start slowly – only take on a few students at a time – don't rush to build an empire overnight, or you can get swamped and this will end in stagnation. Perhaps more importantly, don't take weak students just for the sake of having a bigger group – it will backfire and make more problems. In terms of your research, focus your work in an area that is hot and will lead to good results, and continue to be creative and develop new ideas and new collaborators. My remaining advice relates to people management. In my view, it is important to get the group harmonised and working together, rather than as individuals. This can be achieved by having regular meetings with students/postdocs one-on-one and a weekly group meeting where everyone can discuss their research in an open forum. Importantly, because the age gap between new PIs and students/ postdocs is often small, you need to earn their respect early – act like a leader and show the way forward. In particular, you will need to teach students how to write papers and demonstrate how to think like a researcher to solve problems. That said, you will also need to be flexible and understand that each person is different – they may need different management styles to get the best out of them.

And finally, you need to write more grants to get more equipment and more people – and to grow.

Angela Hay is Royal Society University Research Fellow in the Department of Plant Sciences, University of Oxford. Her research interests include developmental genetics and the evolution of developmental mechanisms. A key question in biology is how molecular changes in developmental pathways led to changes in morphology during evolution. Her research aims to identify the genetic differences that underlie divergence in floral structure between two closely related plant species.

Angela writes:

I have been a new PI/IRF for just over 4 years. After finishing my PhD at UC Berkeley in October 2002, I moved to Oxford for a postdoc position. After 1 year, I applied for my own postdoctoral funding and was awarded a Violette and Samuel Glasstone Fellowship, which I started in October 2004. The following year I applied for an independent research position and was awarded a Royal Society University Research Fellowship, which I started in October 2006. In addition to my fellowship, I have a research council grant that started in October 2010. This pays for two postdocs: a 3-year and a 1-year position, and these people have just started in the lab. I also have a postdoc with her own 3-year fellowship funding who started in October 2009. I have one graduate student who has a 4-year studentship and she aims to submit her thesis in 2012.

For me, the most difficult part of the transition from postdoc to independent researcher was realising that there is never enough time for the tasks required of you! As a postdoc, you work very hard, but the only focus is your own research. As a PI, you have responsibilities to your lab, department, the wider scientific community and often to partners and children. There are concrete deadlines for each of these responsibilities – grant deadlines, teaching and committee timetables, review deadlines, dates for meetings and invited talks and family dinner times! In fact, it is your research that lacks a definitive deadline, and it often takes the major hit when time is short. Prioritising the tasks that are most important for your research is also more difficult. As a postdoc, these tasks are mostly experiments. As a PI, these tasks are publishing papers, securing grant funding and attracting good people. Each of these tasks is difficult, each is a high priority and each needs to be repeated successfully without lapse throughout your career. It is particularly difficult to prioritise these tasks while still performing experimental work as a new PI.

I would advise people applying for fellowships/lectureships to focus on publishing their work – it is one of the few things that you have control over in this process. It is much safer to launch into the unknown with a solid publication record behind you. Why should a selection committee believe in an exciting proposal if you have no track record in delivering this type of research? Think about timing your job/fellowship search in order to utilise your postdoctoral work most effectively for yourself. An ideal scenario is to have a big publication preceding your applications/job talks and then another from your postdoc work within a year of starting your job or fellowship. This second publication takes the pressure off your new research, which may have a slow start.

As a new PI, use the chance to determine your own research direction to the max! Make sure that you are doing something so exciting that you get out of bed each morning just to see what happens next. This goes a long way to keeping you and your group happy. Also, I can't emphasise enough the importance of publishing and the importance of selecting good people to work with. The clock is ticking all the time and any approach you take as a new PI has an element of gamble. For example, time spent securing funding is only a good investment if excellent people are hired. If not, research produced by your own experimental work is more likely to yield publications. On the other hand, good people in the lab accelerate your research output and allow you to concentrate on publishing papers, attracting grants and more good people. There is not a simple solution for success, but I advise new PIs to keep a constant eye on these two things and be ready to change tack to ensure timely publications are produced and good hires are made.

Other than the fact that both Jamie and Angela were awarded Glasstone and then Royal Society Fellowships (the latter but not the former was known by us when we asked them to contribute), their research interests and programmes have virtually nothing in common. However, their experiences and advice are very similar. They both stress the importance of publishing as much as possible as a postdoc (yet as a postdoc, writing nearly always gets put aside in favour of that 'crucial experiment'!). They also both stress that it is really important to hire good people and to manage them well. But perhaps surprisingly to some people, they both suggest that time is precious and limited, despite the fact that they are 'only' doing research. What does that imply for the transition from postdoc directly into an academic job?

New lecturers

One of the probationary lecturers we approached was too busy to answer our questions – however, his advice was – get grants, get grants, get grants!

Pause for thought: How do you interpret the 'too busy' scenario above?

Another lecturer had been appointed to a position which had a year's overlap with the outgoing post-holder. This situation allows for a more gradual transition into the job, as the existing post-holder can share some of the responsibilities. Ten months in, **Chris** writes:

> Over the last 10 months I have been balancing my time mainly between research and teaching. I have also been involved in outreach activities, something that I had not experienced as a postdoc. Thus far, I have not been given any significant administrative responsibilities, although this will change in the future.
>
> Although I was obviously carrying out research as a postdoc, as a PI, the word research carries much greater expectations. I still need to deliver the same outputs (publications, conference presentations, seminars, etc.) that I did as a postdoc. However, there are several other important new objectives that need to be squeezed into the same amount of time. These include generating research income, and supervising PhD students. I have a mentor to advise me, but the responsibility for delivering research is mine alone, and my initial objective is to build up a research group. To do this, I need funding and so my most important objective at the moment is grant writing.
>
> When I was hired, I was told that, in my first year, I was expected to deliver a course of 20 lectures accompanied by 5 hours of tutorials. This did not seem a big deal to me because as a graduate student and postdoc I had previously taught at three different universities in three different countries. However, those previous experiences had been isolated lectures, classes and labs rather than a whole course. Over the last year, I severely underestimated the administrative load that accompanies assessment of a lecture course, i.e. putting together and marking exam papers and assignments. Furthermore, because I now teach engineering students instead of physics students, I have had to learn a new style of teaching. However, next year will be much easier

and, because my teaching load is only going to increase slowly over the next couple of years, I should have enough time to focus on my research.

Although I could spend every hour of every day working (and still not get everything done), I need a balance between time spent on professional and personal activities. Most of my weekday evenings are spent with friends, either doing sport or just socialising. However, I still spend some weekends working. I know people who work more than I do, but I really don't know how they manage it!

Chris's story highlights the fact that gaining teaching experience as a postdoc or IRF will prepare you mentally for your teaching role as a PI. However, just as the word research means different things to a postdoc and a PI, so does the word teaching. In both cases, you need to be aware that there are additional expectations and responsibilities as a PI. Teaching even a short lecture course will be a major time-sink the first time around and, because it is time-tabled, and students are going to show up, you will always make it your priority. Given that this is the case, you need to plan carefully to ring-fence time for your research.

And finally, to show that persistence and hard work really do pay off, we hear **Alina**'s story of a non-traditional route into a PI position:

My experience of gaining independence is somewhat different from that of most people. Not long ago I was a senior postdoc in a research group at a prestigious UK university. The group was reasonably large with three to five postdocs, two to three PhD students, two MSc students and several third-year undergraduate project students at any one time. Things went very well for several years until, due to unfortunate events within the group, the number of papers being publishing reduced dramatically. This had the knock-on effect of reducing the number of grants awarded to the PI and thus a once thriving research group effectively collapsed.

Having had a couple of short career development fellowships, I decided that it was time to move on from postdoc life. I knew I wanted to do research and so there were two options for the future: a scientific career in industry or an academic position at a university. I had done a lot of teaching by this time and I knew that I wanted the independence of choosing my own research projects, so I decided that academia would suit me better than life in industry. My first application for a lecturer position was unsuccessful and, as I had not even been considered for an interview, I was dispirited for a while. Looking around for alternatives, I applied for a fixed-term position of Educational Project Officer, working closely with the Divisional Academic Advisor. I reasoned that this job would give me more insight into academic life and would help me learn how best to tackle applications for academic positions. This proved to be a very good move – at the end of the fixed-term period, I applied for four lecturer positions and was selected for interview for all four. My third interview was followed by an offer of a temporary position providing maternity cover.

Starting as a lecturer is never easy, but when you start by covering for someone else and are immediately immersed in lots of teaching, with little guidance about how the department works or about the teaching ethos, it is a real challenge. In my first term I had a full teaching load, teaching topics that I had never taught before. It was extremely demanding and all of my other commitments had to be put on hold for months. At the end of the second term, I was really and truly exhausted. My temporary position did not involve research, so unfortunately that was put on hold also, no matter how much I missed it. However, my hard work seemed to be appreciated, because before the end of my contract, the department offered me an extension with more responsibilities, including research and the potential for being made permanent.

And so, this is my story of the postdoc to PI transition. It was hard work and it is not the way you would normally choose to do it; however, it worked for me! What would my advice be for new PIs? Be aware that the transition is more difficult than you imagined, prioritise things and keep in mind that you will have to make compromises. Take every step one at a time and, whatever you do, do it right. Do not compromise on quality, even if you compromise on quantity. Be open minded, flexible and take everything that comes along with humour. Try to balance your teaching and research – *this* is something I need to do now! And take time to relax from time to time – you need it.

As you read and reflect on these stories, several interesting aspects of the transition from postdoc to PI emerge. We have already mentioned one – a different approach to time management. Other aspects can be viewed almost as a mirror image of what you have done as a postdoc – but now you are on the other side of the fence. It is often at this stage that timely advice from your mentor, your head of department or a good colleague can save you a lot of time and hassle. There are also printed and electronic resources that are particularly useful to consult when you are first starting out (e.g. Barker, 2002).

How we did it

In our workshops we invited new lecturers and current IRF holders to come and give short presentations about their experiences of the transition from postdoc to PI. Presentations were normally followed by lively question and answer sessions.

Summary

How individual academics settle into their new roles is of particular concern both to them and to their heads of department. The existence (or not) of a collegiate atmosphere in the department, quality induction procedures and appropriate administrative support can make (or break) their careers (Smith, 2010). There are several common factors identified by young PIs that mark the transition period from being a postdoc. Perhaps the most difficult one to manage is the reduced flexibility in terms of when things have to be done, and the increase in the number of things that need to be done. This is amplified by an increased responsibility for your own and other people's careers. The key focus at this career stage is securing funding for research, and the advice given is to make sure that you have a solid publication record to back up your applications. There is no doubt that the postdoc–PI transition period is exciting and challenging, but it is also probably the most distorted in terms of work–life balance compared with being a postdoc and with later years.

Selected reading

If you only have time to read one book, make it:
Barker, K. (2002). *At the Helm: A Laboratory Navigator.* New York: Cold Spring Harbor Laboratory Press.

Other texts:
Smith, J. (2010). Forging identities: the experiences of probationary lecturers in the UK. *Studies in Higher Education*, 35 577–91.

12

Managing people

There are so many theories of how best to manage people that whole courses are run on this topic alone. Here we advocate thinking 'win–win' and 'seeking first to understand' in order to create synergy. In practical terms, we provide a number of scenarios that require a management solution. These are managing up (your bosses), sideways (your peers) and down (your research group). We recognise that different management skills are required to handle different situations; however, the 'win–win, seek first to understand and synergise' approach underlies all cases.

The theory

Managing people is an art not a science and some would even say a black art. There is an apocryphal saying that managing academics is like herding cats or pushing a wheelbarrow full of frogs. However, you only have to look at the success of 'big science' projects such as the Large Hadron Collider to realise that this saying is only partially true.

Managing people in academia, or more generally in research, does not necessarily mean working within well-defined hierarchical structures, such as those found in industry. However, there is a common code of good practice, which underlies managing others. It requires recognition of the fact that we are *interdependent* on others for achieving outstanding results. In the context of Covey's seven habits of effective people (Chapter 1), a good group leader would think 'win–win' (*habit 4*), seek first to understand (*habit 5*) and synergise (*habit 6*) (Covey, 2004). What does this mean?

Briefly, thinking 'win–win' requires us to think in terms of mutual benefit and to seek solutions for conflicts and problems in that context. In order to be successful, it has to be based on the philosophy of cooperation, not competition. As such, it may be hard to accept and apply in cases where being the first to discover something matters a lot. However, we need to distinguish between external and internal competition. What we discuss in this chapter applies only to managing relationships within a group or department. In this case, good cooperation can make the difference between success and failure. Notably, cooperation requires maturity (courage and consideration). Seek first to understand requires *really* listening to another person first, and then explaining your point of view clearly and succinctly so that both parties can develop mutual trust. The sixth habit of synergy is the most elusive and the most difficult to develop – but when it works, it delivers great results. Synergy builds on differences between people that are recognised, acknowledged and used creatively to achieve more than would be possible if individuals were working alone. In this chapter we concentrate on win–win and synergy; seeking first to understand is discussed more in depth in Chapters 13 and 14.

The practice

In terms of social structure, we are always managing on three levels: *managing up* (your bosses), *managing down* (your group) and *managing sideways* (your peers). What we discuss here is not a technique; we are proposing an approach consistent with your personal and professional roles and integrity. The approach requires an open mind, the ability to sometimes resist outcomes that provide immediate gain and a willingness to seek solutions that are mutually beneficial.

Managing 'up'

The lower you are on the career ladder, the more likely it is that jobs will be delegated to you. Depending on the structure of your institution you may have very little room for manoeuvre. Consider the examples in Box 12.1.

Box 12.1
Managing 'up' scenarios

1. Your head of department (HoD) asks you to a give a course of four lectures to first-year undergraduates next year. The lecture topic is well beyond your immediate expertise and the course will take you at least 2 weeks to prepare. You do not feel that you have the time to do it, nor do you feel you will gain much from it.
2. Your HoD asks you to take on an administrative role. It is an important job and you know you could do it well. However, you already have a greater administrative load than most of your peers. If you don't do the job, someone else will have to and they may not do it as well. If the job is not done well, it will have a knock-on effect on other things you do and will disrupt some of your work.

Pause for thought: What would you do in the first case? How would you handle the second case?

Our analysis: These are open-ended problems and the outcomes could well depend on the personalities involved. However, we think that the first thing to do in both cases is to prepare carefully before talking to your boss. If asked on the spot to take on a new duty, you could say that you need to check your other commitments before discussing it fully.

In the first case there may be a colleague who is better qualified to teach the new course and maybe you could take over one of their courses in exchange. You could suggest this as a compromise; however, your head of department may have a longer-term view on your career and duties and may not want to release you from this particular job. In that situation you will have to convince yourself that there is never a good time to broaden your horizons – so you may as well do it now.

In the second case you need to decide which of your current duties could be taken on by someone else, and convince your HoD that, in order to do your new job well, you have to give up something. This is in the spirit of 'don't bring me the problem, bring me the solution'. However, you should avoid naming colleagues – your boss knows the administrative load of his or her staff. In this situation your HoD has a vested interest in the department running smoothly, and is likely to listen and negotiate for the best – in his or her view – outcome. So this may well be a win–win situation. If you go for 'no deal', then somebody else will have to do the job and it may have an adverse effect on your work.

Pause for thought: Think about a recent situation where you feel that you 'lost'. Do you think the other party 'won'? If so, how could it have been handled differently so that both sides won?

Managing 'down'

In this situation you are in charge and have control over the process although not necessarily over the outcome. It is our experience, supported by countless case studies in management, that leading by personal example is a very powerful motivation for others. In Box 12.2 we present some typical examples:

Box 12.2
Managing 'down' scenarios

1. Your first-year PhD student is very enthusiastic and wants to try many different things. If allowed to do so, s/he is unlikely to have anything coherent to present at the end of the first-year. This will influence the outcome of their first-year assessment.
2. Your graduate student is having difficulty writing his/her thesis. Lab work finished 3 months ago but you have only seen one Results chapter. There are only 6 months left before the 4-year deadline for completion.

3. You have someone in your group who consistently leaves communal areas untidy.
4. You have a group member (person 1) who is very experienced with technique A. Another group member (person 2) has had a paper almost ready to submit for 6 months. The paper is waiting for one more experiment that requires technique A. Person 2 can't get it to work. Person 1 is willing to do the experiment but person 2 doesn't want to dilute the author list on their paper. It would take person 1 less than a week to do the outstanding experiment.

Pause for thought: Consider the first two problems. How could you steer your student's motivation?

Our analysis: In both cases you need to consider the underlying expectations of your students. In the first case, you could agree on the most important tasks to be completed by the end of the year and then as long as they are finished, allow the student to try other things as they wish. The problem here is that, if the student falls behind schedule, you may have to enforce the solution or at least renegotiate it. This may take more of your time than you wish, but keeping a motivated student on track is worthwhile. In the second case you need to establish the source of the problem. It may be writer's block but there may be other factors that you are unaware of. You may have to help by writing a detailed work breakdown schedule for the thesis (see Chapter 6), and then get the student to decide the order and timing of chapter delivery.

Pause for thought: How would you encourage your group to keep communal spaces tidy?

Our analysis: There are many ways of dealing with this problem and there is no generic solution. Keeping the work environment tidy is very difficult if people do not feel responsible for it. So it may be that you need to talk to the whole group and divide duties equally between the members, with a clear schedule of what needs to be done by when and by whom. Contributing to the schedule yourself will not only show that the responsibility is truly shared, but will also show what standard of tidiness you expect. The principle here is that all group members are equally valuable and equally responsible, so your treatment is fair. Some groups may be able to interact smoothly and resolve such conflicts before they get out of hand; however, you cannot rule out the need for your intervention at some stage.

Pause for thought: How would you convince person 2 in scenario 4 that it is better to get the paper published as soon as possible?

Our analysis: This particular problem calls for a synergistic solution as person 1 brings a bigger contribution than just doing an experiment – the paper becomes publishable. You would need to get person 2 to think beyond this one paper, to set it in the context of his or her career. Is there scope for the work to be published in a higher-quality journal if it is published sooner? Is there anything that person 1 could offer in terms of joint authorship on another project? There may be other approaches, depending on all of the personalities involved. However, the bottom line is that an unpublished paper is a waste and serves nobody.

Managing 'sideways'

In this case, you and your colleagues are equal so there is no power base to act from (even if you wanted to). Seeking win–win solutions is thus the best solution as an outcome agreeable to all parties ensures a good working atmosphere and has far-reaching consequences. Your reputation as a good colleague and a good citizen may also be just what is needed for promotion or for landing a desirable job somewhere else. Typical situations that you may encounter are shown in Box 12.3.

Box 12.3
Managing 'sideways' scenarios

1. You are Director of Graduate Studies. A student comes to you with concerns about inadequate supervision by one of your colleagues.
2. Your group shares a greenhouse with another group. The PI of the other group is not as concerned about pest infections and thus their plant care practices are not as stringent as yours. This leads to infection of the whole greenhouse.

Pause for thought: How would you proceed in the first case?

Our analysis: Clearly, this situation calls for a lot of diplomacy because the student's supervisor is your equal in terms of the university's teaching and learning hierarchy. You will first need to establish facts, find out what the student means by 'inadequate supervision' and how the supervisor sees the situation. Your main role is to listen empathetically and to get the two sides to try and resolve the situation by listening and talking to each other.

How we did it

In preparation for this seminar, we sent out a list of various management scenarios and asked participants to come to the seminar prepared to discuss how they would resolve each situation. At the end of the seminar, we collected the participants' conclusions and opinions into a summary. Each group of participants came up with conclusions similar to those outlined in the Summary.

Summary

Management solutions are always personality dependent. As such, when managing people, it is always good to discuss expectations on a regular basis. When problems arise, open-minded and frequent communication with all concerned parties is essential. Managing 'up' requires you to influence the situation to get the best outcome for you, whilst still satisfying your manager. In contrast, managing 'sideways' requires diplomacy and a desire on both sides for a win–win solution. In your role as PI, you will have to deal with all the people in your research group on an equal basis and will need to consider their needs in addition to your own. Key points to remember are that (1) it is your responsibility to manage conflict within the group – you cannot always expect them to resolve it themselves because they are all managing sideways and (2) people work in different ways – when assessing others, consider outputs not the method of input.

Selected reading

If you only have time to read one book, make it:
Covey, S. R. (2004). *The 7 Habits of Highly Effective People*. London: Simon & Schuster.

Other text:
Mullins, L. J. (2010). *Management and Organisational Behaviour*. London: Pearson Education.

13

Building a research group I: doctoral students

At some point in your career, you will think back to how you supervised your first graduate student and will probably feel very grateful to the student concerned for his/her tolerance. Your methods of graduate supervision will constantly evolve and, in an ideal world, your supervisory skills would continue to improve over the course of your career. In reality, they will probably peak well before the end. Regardless, you have to start somewhere. In this chapter we look at things to be considered when recruiting and supervising graduate students.

The theory

You are now finally recognised by the university and your peers as a PI. One of your first tasks is to build a research group. You might have previously supervised project and PhD students; however, as an official supervisor you have a lot more responsibility. What is a PhD about? In the UK it is typically 3 to 4 years duration, in Europe 4 or 5 years, in the US often longer. In all cases a PhD degree is awarded 'for an original contribution to knowledge'. Educational research indicates that students may show originality in a number of ways (Box 13.1) (Phillips & Pugh, 2005).

As a supervisor it is your job to ensure that your students have the opportunity to demonstrate their contributions to knowledge.

Pause for thought: How could the research you did for your PhD be described in terms of the statements in Box 13.1?

The practice

When admitting graduate students to your research group, you need to consider five things: funding, scope of project, student selection, supervision and work environment. We look at each in turn.

Box 13.1

Examples of original PhD contributions

- Setting down a major piece of new information in writing for the first time.
- Continuing a previously original piece of work.
- Carrying out original work designed by the supervisor.
- Providing a single original technique, observation or result in an otherwise unoriginal but competent piece of research.
- Having many original ideas, methods and interpretations all performed by others under the direction of the postgraduate.
- Showing originality in testing somebody else's idea.
- Carrying out empirical work that hasn't been done before.
- Making a synthesis that hasn't been made before.
- Using already known material but with a new interpretation.
- Trying out something in the UK that has previously only been done abroad.
- Taking a particular technique and applying it in a new area.
- Bringing new evidence to bear on an old issue.
- Being cross-disciplinary and using different methodologies.
- Looking at areas that people in the discipline haven't looked at before.
- Adding to knowledge in a way that hasn't been done before.

Funding

PhD studentships can be funded in a number of ways: through specific applications to funding bodies, on a research grant, by industry, by departmental/university programmes or by the students themselves. In all cases, there needs to be a maintenance grant and fees. In some cases, adequate research costs are also provided, but in many cases they are not and you need to find another source of funds to subsidise the project.

Project scope

In most cases you will define the research field and project scope for a student. To do this, you need to think carefully about the facilities available and you have to bear in mind that a PhD student is not going to be as fast or as efficient as you would be. There is a steep learning curve even if you choose somebody with experience of a particular technique and/or with good background knowledge. Hence it is important to make sure that the student will not be working on your cutting-edge project or the likelihood is that you will end up frustrated by slow progress, and maybe scooped by competitors. In deciding the project scope, it helps to talk to your colleagues and your mentor, and to have a look at PhD theses that have been successfully submitted in your department. Remember that your students will expect you to have specialist knowledge in the field they are working in.

Student selection

Once you have secured funding and have an idea of project scope, you will need to select a student. In many institutions, particularly in the US, students enter a general graduate programme and then do 'rotations' through different research groups. This provides time for both you and the student to assess whether you can work together. However, in other cases, you need to select a student on the basis of applications in response to an advertisement. In this case you will need to interview prospective students – either face-to-face or by video-link or telephone. The interview is essential as the supervisor–student relationship is very personal and therefore as a supervisor, one of your main concerns at interview should be whether you could work with the candidate. The interview can either be formal or informal but should always involve two or three interviewers. For a formal interview, make a list of questions and divide them between the panel members.

Pause for thought: Think back to your PhD. Did you get on well with your supervisor? What was particularly good for you? If you didn't, why did it not work? What would you have liked to be different? Write down a few pointers.

During the selection process it is important to remember the following:

- For many students, the PhD interview will be the first interview they have ever had. This means that even really bright students may feel intimidated and so you should work hard to put them at ease. If you don't, you will not get the best out of them. A bottle of water on the table (and a glass!) is courteous and often essential if you want them to be able to speak. Make sure to introduce all members of the interview panel to the candidate. It often helps to tell the candidate at the very beginning, how the interview is going to proceed. A general question about how they have got to the point of applying for the position also allows them to start talking without having to think really hard about the answers.
- Whilst the CV gives you information about 'what' the candidate has done, it will not tell you 'why'. For example, someone may have on their CV that they have spent time reading to a blind person. In reality, this could either be because s/he is a caring individual or because s/he had to do it as community service following prosecution in a young offenders' court. The 'why' is what distinguishes the individual. As you want to gain as much knowledge as possible about the candidate, 'why' questions are a good thing.
- Open-ended questions allow the candidate to tell you things that you might not have thought to ask (see example in Box 13.2).
- If the candidate claims to be good at something, ask for an example of a situation where they have demonstrated that ability.

Box 13.2
Sample interview questions for PhD candidates

1. Why do you want to do a PhD?
2. What personal traits do you think you need to complete a PhD?
3. Can you give me an example of a situation that you have been in that demonstrates that you are persistent?
4. What do you expect of a PhD supervisor?

- Make sure that the candidate has realistic expectations of what a PhD entails and check what their expectations are of you and the system.
- Check whether the candidate has both general and specialist knowledge of the area. Maybe you won't mind if they only have one or the other, but you need to know which area will need improvement.
- Give the candidate time to think about their answers. Short periods of silence are not a bad thing.
- Finish by asking whether the candidate has any questions to ask you and also ask them whether there is anything they wished you had asked them but hadn't. The latter, protects you against claims of 'they didn't even explore XYZ so how were they able to judge?'
- Where possible, make sure that the candidate knows ahead of time what the selection criteria are. After the interview, keep a record of how you assessed their ability in relation to each criterion and how they performed.

Pause for thought: Imagine you are going to interview your first student. Write down a few questions you would ask this person.

Supervision

PhD supervision has to accommodate and encourage the growing independence of the student. If you supervise students too strictly, they may never realise their potential; if you leave them to their own devices too early, they may fail. When you agree to supervise a student, you are therefore accepting a major responsibility. If you don't feel able to fulfil this responsibility for the duration of the student's graduate career, for whatever reason, do not accept the student.

So what are a supervisor's responsibilities? Universities and funding agencies have supervisory guidelines that have to be followed and we recommend attending seminars on supervising PhDs organised by your university whenever possible. Clearly, the most important supervisory role is to steer the student towards the successful completion of a PhD. In doing this, you have to remember that the student is starting out on an unknown journey, so you have to map out some milestones. The degree of supervision required will then vary with time – you can reasonably expect that, in the first year, your students will require more frequent meetings than in their final year. You will also have to be tuned into your students' mood – there will be many difficult moments and disappointments along the way, so it is important to be able to provide a different perspective. The key points to bear in mind when supervising are:

- It is up to you to make sure that students are aware of the standard expected of them. This is particularly important with respect to assessment procedures. If they are to meet the required standard, they need to know what it is.
- It is important for expectations on both sides to be aired at the start of the research and to be reiterated at regular intervals because expectations change. It is sometimes worth remembering that your student's view of you is likely to progress from total awe to healthy disrespect over the period of their research.
- It is often helpful to draw up an annual training plan in consultation with the student. This can outline both general and specific research objectives for the year. This plan can be used by both student and supervisor to measure performance in relation to expectations.
- Funding agencies now assert that, in addition to turning out specialists in narrow research fields, supervisors should help students acquire transferable skills. In a nutshell, the future PhD should be able to think independently, be able to apply problem solving skills to new situations and be able to communicate research findings in a variety of ways.
- It is worthwhile encouraging your student to have a second supervisor, even if university guidelines don't insist on it. This arrangement can benefit both you and the student if either of you requires re-enforcement of an idea. (Of course, you can't guarantee that this will always work in your favour!)
- If a project is jointly supervised by two PIs, it is important to clarify the responsibilities of each supervisor and again make sure that the expectations of all three parties are aired.
- You should be available to meet with your student regularly. Your schedule will be much busier than theirs, and it is easy for you to let the weeks slip by without a meeting. Consider time-tabling a weekly meeting, even if you both decide you don't need half of them on the day. At least that way, the student knows that they have a dedicated regular time slot where they can air concerns or get feedback.
- You have to tell the student how well (or otherwise) they are getting on. Without feedback, they are unlikely to improve.
- If the student's project is part of a larger effort in the group, it is your responsibility to ring-fence the student's work. Enthusiastic postdocs often fail to

recognise or respect project boundaries in their effort to find 'the answer'. You must make sure that the student has the opportunity to produce a defendable, independent thesis.

- Research councils, the university and departments all publish documents on student–supervisor responsibilities – make sure you read them and are familiar with the contents. It is also advisable to attend a course on supervising students – most universities hold them for their own faculty.
- The process of doing a PhD is also a process of maturing – from student into a scientist. This process can be emotionally taxing for both student and supervisor.
- If you have considered and acted upon all of the above points, it is important to remember that, if the student fails an assessment (or indeed their thesis), it is not you who has failed. It is often too easy to take such events personally, which is unhelpful both to those carrying out the examining and to the student.

Pause for thought: Are there additional issues, not mentioned above, that you consider important?

Of course, the PhD culminates in writing a thesis. Many students find this a daunting task because they don't think writing is as important as doing experiments and generating data; hence they will postpone writing until the last minute. Obviously, writing a thesis is different from writing a paper and you will have to help the student learn how to do it. One way is to encourage the student to write up draft Results chapters after a series of experiments is finished or to carry out a literature search for part of the Introduction. In this way, by the time the main body of work is fit for submission, there will be plenty of written material ready to be put together and rewritten as appropriate. Whichever way you choose to help, the most important thing to remember is that you need to teach them *how* to write, not *what* to write. You have to be very careful not to rewrite the thesis for them, as this would invalidate their work.

Work environment

While every lab has its own unwritten rules that will become obvious to the new arrival in due time, there are general procedures that have to be explained and enforced in a more formal induction process. These procedures go beyond just keeping your bench organised or cleaning up after yourself – they involve guidelines for interacting with others both within and outside the research group. Importantly, knowing the lab rules and ethics clarifies expectations, and while it may not prevent conflicts it certainly helps you as a PI to solve them. Although it is a time-consuming activity, the importance of such an induction process is thus well recognised. Because things quite often have to be repeated several times before they are internalised, it can help to suggest that students look at a practical guide to life at the bench (e.g. Barker, 2005).

Finally, it is important to recognise that the number of 'non-standard' graduate students is increasing in many institutions. They may be part-time, mature or be non-native English speakers. In these cases you will have to think even harder about how to help and guide them so that they get the best out of their research. For example, part-time students may need guidance in time management, and mature students may need help fitting in with your research group, as they may have family responsibilities and thus less time for socialising. Cultural differences can also present a challenge. For example, Asian students typically look up to a supervisor with greater reverence than other students. This can cause problems when the former are encouraged to present independent ideas that challenge those of their supervisor.

How we did it

In our seminars we set up mock interviews, with participants interviewing current undergraduates for PhD places. As a group, we then analysed the questions asked and the responses given to determine which questions had elicited the most useful responses. We also spent time discussing individual experiences of being supervised as a PhD student. This led to a consensus list of good and bad supervisory practice.

Summary

In this chapter we have discussed the student–supervisor relationship. If you appoint an inappropriate person to your first PhD studentship, it can seriously affect both your science and your morale, not to mention cause possible damage to the student's self-esteem and be a waste of time for both of you. We have therefore outlined important areas for consideration during the recruitment process. While each student needs to be treated on a case-by-case basis, we have also provided some guidelines to help ensure that your relationship with your PhD students is a positive one.

Selected reading

If you only have time to read one book, make it:
Phillips, E. & Pugh, D. S. (2005). *How to Get a PhD: A Handbook for Students and Their Supervisors*. Maidenhead: Open University Press.

Other texts:

Barker, K. (2005). *At the Bench: A Laboratory Navigator.* New York: Cold Spring Harbor Laboratory Press.

Finn, J. A. (2005). *Getting a PhD : An Action Plan to Help Manage Your Research, Your Supervisor and Your Project.* London: Routledge.

14

Building a research group II: recruiting and supervising postdocs

Postdocs are vital for the health and wealth of your research group, not least because most scientific papers have a postdoc as first author. When you recruit someone to a postdoctoral position, it is therefore important to be able to develop a positive working relationship with them. In this chapter we consider how to recruit, supervise, guide and motivate postdocs. In particular, we suggest goals for an 'ideal' supervisor and suggest practical ways to achieve those goals.

The theory

Postdoctoral researchers occupy a transition zone between well-defined PhD study and equally well-defined PI positions. In effect, postdocs can be compared to medieval journeymen – having finished an apprenticeship, they would hone their skills by travelling from one master to another until they were competent enough to become masters themselves (Harris, 2008). But how do modern-day journeymen find a job that pays a salary and has future employment prospects? Most postdocs are funded by personal fellowships or by research grants awarded to PIs. As a PI, you have to consider very carefully who you want to employ on a grant, particularly as it is becoming increasingly difficult to attract research funding. Unlike graduate students who embark on a course of study with the defined endpoint and goal, postdocs embark on a project for many different reasons. Maximising the potential of both your research programme and of your postdocs' careers (be they ultimately in science or not) requires that you recognise what these reasons are.

The practice

Recruitment

There are a number of reasons why people decide to carry out postdoctoral research – for example:

1. They are committed to a career in academia and this is the next step towards becoming a PI.
2. They are interested in research, but are not yet sure whether they want to stay in academia or move across to industry. They see a postdoc in academia as less likely to close doors at this stage.
3. They know they don't want to stay in scientific research but recognise that scientific skills are important. They want to hone these skills further before doing something different.
4. They don't really know what they want to do, but they are good experimentalists. A postdoc is thus a job to fill the time and pay the bills while they think of an alternative career.

Depending on an individual's motivation for doing postdoctoral research, they will have different needs and expectations. You will also have different expectations, depending on the type of project that needs to be carried out. For example, do you want someone who will challenge you and who will take the work in directions you might not think of, or do you want someone to complete a defined piece of work that requires a certain skill? Either way, it is crucial to clarify expectations on both sides before you hire someone and you therefore need to ask penetrating questions at interview (see examples in Box 14.1).

Box 14.1
Examples of questions to ask when recruiting postdocs

- Why did you apply for a postdoc?
- Why did you choose to apply for this particular project?
- What did you contribute to the papers you have published?
- Where do you see yourself when this grant ends?
- Do you like working with PhDs? Undergraduates?

Pause for thought: Write down other questions that you would ask a prospective postdoc.

Supervision

Once you have decided that someone is the best fit for your project, you have to consider your responsibilities and the influence that you will have on their career. Obviously, both of you will want the project to end with the best possible outcome. However, your views on what is 'best' may differ. Here we need to signal that, because you are your postdoc's line manager, your relationship with them is not mentoring (for information on the latter, see Chapter 19). Your responsibility is

---

Box 14.2
An ideal postdoc advisor

- Provides the right amount of intellectual space (some postdocs need more than others).
- Provides physical space and resources to enable the work to be carried out efficiently.
- Challenges intellectually.
- Enforces scientific rigour.
- Promotes the postdoc's work nationally and internationally.
- Protects the postdoc from unnecessary administration and bureaucracy.
- Helps the postdoc to develop their career.

---

greater, and it is not advisable to become good friends, as it may be difficult to enforce the discipline of scientific rigour. Importantly, you may find that you change in the face of such responsibility, and you may need to develop new ways of thinking and acting. Guidelines for advising are shown in Box 14.2. The need to guide postdocs and other research staff in a structured, yet not stifling, way has been recognised by funding bodies such as RCUK and there are good practice guidelines that have been put together, such as the Concordat to Support the Career Development of Researchers. UK universities, many research institutions and professional bodies are signatories of this Concordat (http://www.researchconcordat.ac.uk/). These institutions organise courses and seminar series such as ours, which provide opportunities for individual career development.

On a more personal level, some practical ways to achieve the goals listed in Box 14.2 include:

- Have regular meetings to discuss scientific progress (timing should depend on the amount of intellectual space the postdoc needs, but should probably be at least once a month). Be sure to provide both positive and negative feedback. People can't improve if they don't know where they are going wrong.
- Provide a forum to encourage and train people to keep on top of the literature.
- Encourage postdocs to attend conferences and give talks. If it is not possible for them to give a talk, get them to take a poster to a small meeting.
- Introduce postdocs to other people in the field.
- Help devise an annual personal development plan that addresses both scientific and personal goals. Use the plan to maximise transferable skills training for those who do not want to stay in research, and to develop teaching and administration skills for those who do.

Pause for thought: If you have postdocs working with you, which of the above do you already do? If you plan to recruit a postdoc, which of the above points would you make your priority, and why?

Guidance

Guidance is not the same as supervision as it goes beyond the practical issues discussed above. This is more about how to do science and to develop and retain professional integrity during the process. The career pressure on a postdoc to publish is great, as it is on you to deliver your project on time with good results. In your regular meetings you need to emphasise the requirement to record data in a way that can be followed by someone else. You should also encourage your postdocs to develop a critical mind, emphasise that the foundation of good science is reproducible results and stress that there is a fine line between harmlessly collating the best results and shifting data points to better fit a theoretical curve (Goodstein, 2010). They should leave your group knowing that it is OK to challenge established authorities, that unexpected results cannot be suppressed and that Adobe Photoshop is not the answer to good data presentation.

Pause for thought: How can you encourage others to develop a critical mind?

Maintaining motivation

A good project, sufficient money and enthusiastic competent postdocs are key ingredients for success. But we need to add an essential extra to this mix, namely group motivation. In science, this topic is hardly ever raised – we are supposed to be motivated and to magically maintain momentum day in, day out. Clearly untrue. As a PI, you not only need to keep yourself motivated, but you also need to motivate others. So, how do you do it?

Generally speaking, the social conditions in which human beings function makes them either active and engaged or passive and alienated (Ryan & Deci, 2000). Studies have shown that members of a group perform best when they can work in a self-determined way, and the three conditions for self-determined behaviour are: *competence*, *autonomy* and *social connectedness* (Ryan & Deci, 2000). The context in which this works for research groups has been illustrated by Alon (2010).

Gauging the competence of group members is a necessary task because competence is the first prerequisite for motivation. If a task is too difficult, your postdoc will become demotivated; if it is too easy, he or she will become bored – you have to aim for peak performance or 'flow' in psychological terms.

Pause for thought: What level of competence do you expect of a postdoc (current or prospective) on your current project?

The second essential component of motivation is autonomy. This is the amount of intellectual space needed for a person to develop a sense of ownership of the work.

This is a carrot rather than a stick approach and may require a lot of patience on your behalf – often it is much faster to solve a problem yourself than wait for a postdoc to do it. The amount of autonomy granted to each of your group members will vary across the spectrum between management and freedom – hopefully, it will not touch on micromanagement or neglect. In ideal conditions it will become self-regulatory as the group members mature.

> **Pause for thought:** Think back to your first postdoc or to your time as a PhD student. Were you given enough autonomy to stay motivated through the project? If not, what would you have liked to be different?

Finally, motivation depends on social connectedness: does anybody care how well I am doing? Do I care how others in my group are doing? Can anybody give me help when I need it? Can I help others when they need it? The best performing research groups have a high degree of social connectedness. One of us was a postdoc in a group that had regular Friday lunches in a pub, often with another group working in a similar area. A couple of lazy hours spent socialising were amply repaid with free-flowing discussions and the knowledge of whom to turn to with a particular problem.

> **Pause for thought:** What are you already doing to bring about social connectedness in your group? How could you interact socially with your group without compromising your professional relationship?

How we did it

In our seminars, participants came prepared to discuss their own experiences of being a postdoc. Perceived good and bad practice in terms of the postdoc–supervisor relationship were discussed and a consensus generated.

Summary

Postdocs are vital to your research and to your reputation as a PI. In this chapter we have discussed recruitment, guidance and supervision, and how to build a good community spirit in your research group. There is no doubt that managing postdocs can be challenging and at times frustrating; however, the effort is nearly always worthwhile. Positive postdoc–supervisor interactions can lead to long-lasting relationships of mutual respect, trust and shared scientific interests.

Selected reading

If you only have time to read one book, make it:
Goodstein, D. L. (2010). *On Fact and Fraud: Cautionary Tales From the Front Lines of Science*. Princeton: Princeton University Press.

Other texts:
Alon, U. (2010). How to build a motivated research group. *Molecular Cell*, 37, 151–2.
Ryan, R. M. & Deci, E. L. (2000). Self-determination theory and the facilitation of intrinsic motivation, social development, and well-being. *American Psychologist*, 55, 68–78.

15

Interacting with others

Over the course of your career, your science will be enhanced through interactions with others, both in your own field and beyond. Our accomplishments are noticed and appreciated by others, but we also need to notice and appreciate the achievements of others. In academia such interdependence is seen in networking, collaborations and in 'community service'. In this chapter we explore the benefits and costs of these three endeavours.

The theory

Covey's philosophy for success (see Chapter 1) is focused on developing seven 'habits', of which three relate to how we deal with others. Since we act upon, and are acted on, by the world around us, it is essential to identify the limits of our influence. In this context there is a *Circle of Concern* that comprises events and actions that we have no real influence over, but which still affect us, and a *Circle of Influence* that comprises events and actions that we have some control over (Covey *et al.*, 1994). The art is to focus our efforts on the latter (note that proactive people ensure that the circle of concern is encompassed by the circle of influence).

Covey's main audience is managers and industrialists – can scientists, and in particular academics, benefit in any way from his philosophy? One of the main aims of science is the production of knowledge (Ziman, 2000) and on this premise many similarities can be found. For example, scientists are promoted on the basis of the quality and number of papers published (output), the total amount of grant funding awarded (resources necessary for production) and on their national and international reputation (brand name). More importantly, the success of Covey's philosophy is in large part due to a combination of ethical principles, self-development and action in the real world. In his approach 'fundamental principles' such as fairness, integrity and honesty form the cornerstone of behaviour (Covey, 2004). It may be no coincidence that each of these principles is central to the scientific process.

In considering the link between Covey's tenets of independence and interdependence, the key thing is that we can achieve more in 'exchange' with others. In academia this exchange can be considered in the context of networking, collaboration and 'community service', each of which is considered below.

The practice

Networking

The purpose of networking is to establish lasting relationships that will enable you to carry out your research in an international context – it is essential – not only is it a way to get noticed and be remembered, but it can also be a preamble to successful collaborations. So how do you do it? In its simplest form, networking at a conference is analogous to speed dating – you have a very limited time to present your research and to make an impression before you (or your audience) moves on to the next person. So, you need to listen carefully, but you must also be able to convey the importance and excitement of your research in short sound bites – nobody likes to listen to a bore. A related art is to know how to move on gracefully without offending, particularly if you spot someone that you must, or want to, talk to. Of course, you can also network 'remotely' via e-mail, phone or video-conferences but each of these forums has their own etiquette and are often more difficult to initiate than face-to-face networking.

Why network?

- People are more likely to remember and respect your science if they remember and respect you.
- When people are asked to recommend someone to give a talk or write a review (or anything else), they need to know that they can trust the person nominated to do a good job.

How to network

- Discuss your work with seminar speakers who visit the department, even if they are not working in your field.
- Ask questions after seminars.
- Present posters at meetings and make sure that you explain your work to as many people as possible.
- Give talks at meetings – even if they are short talks at small meetings – the more the better.
- Target people at meetings whom you know work in your area. Ask about their work and explain your work to them.
- Go on a self-invited seminar tour.

Bottom line

Get noticed – for the right reasons!

Collaborating

Collaboration affects everybody. There are projects that are simply impossible to do on your own. Large Hadron Collider experiments, for example, involve hundreds of scientists. There are other subjects such as molecular modelling where projects can, in principle, be carried out by individuals, but the effort required to model a complex piece of matter would quickly leave a loner far behind large groups working in the same area. And there are studies where it is best to work on your own, at least initially – you can develop an idea and then decide how to pursue it. When you do collaborate, however, it is always most effective when synergies are developed between partners – and synergies arise when we consistently use a win–win strategy and seek first to understand (Chapter 1; Covey *et al.*, 1994).

Why collaborate?

- To move into a new field where you need someone else's help.
- To help solve someone else's problems.
- To get materials from someone else.
- To tackle an interdisciplinary problem.
- To avoid competing with someone.

What is needed for a successful collaboration?

- Good communication between all parties at all stages.
- Clear understanding of expectations on all sides.
- Equal commitment from all parties.
- Shared enthusiasm to see the project through.

Possible gains from collaborating

- Intellectual stimulation.
- Fast progress.
- Fresh insight to a project.
- Publishing back to back instead of getting scooped!

Possible problems of collaborating

- Closure can be difficult because nobody feels ultimate responsibility.
- Establishing your role/identity in the project, particularly if you are a junior PI, can be hard.
- Sorting out author lists on papers can be complex and emotional.

- Different expectations/standards about quality and interpretation of data can lead to serious disagreements (in the worst case to accusations of fraud).
- Lack of understanding about different areas can mean you are unable to critically assess data produced by a collaborator.
- The project will need consistent attention to keep things on track.
- International collaborations or different nationalities working in the same lab can lead to different expectations that are assumed rather than articulated.

Bottom line

Collaborations can be beneficial and fun but only if they are managed properly. Know why you are collaborating and what you expect to get out of it before you start – don't just drift into it (see case study in Box 15.1).

Box 15.1
Case study 1

Emily has been doing quite well on her project but her supervisor is leaving the university. She has been asked to join another group and extend her field of study. This would be a more interdisciplinary area involving computer modelling, an area that she has been interested in but has no experience of. She sees it as a new challenge and is looking forward to joining the new group.

The new PI is rather inexperienced; however, he already has a postdoc who is using the programme that Emily has to use. The PI asks the postdoc (Hank) to train Emily, but he is not willing to do it. He uses every possible excuse not to help and she is therefore not able to make much progress. Emily is increasingly concerned about the lack of reliable results from the programme and also by the amount of time that is being wasted. She suggests to her boss that she gets help from people in another group. However, the PI does not want this because he thinks it will reflect badly on him – he insists that Hank is the only person who can work with Emily on this project. As Hank is still not helping, the PI proposes that both Hank and Emily can be co-authors on a conference paper with him, on condition that Hank helps Emily. Emily has some concerns about this agreement, and sure enough Hank disappears on vacation for a month without a word to the PI or her.

As the abstract has been submitted to the conference, and Hank is nowhere to be seen, the PI reluctantly agrees that Emily can seek help elsewhere, promising that Hank will be more helpful in the future. However, after Hank's return, all hell breaks loose – he is furious with Emily for 'tampering with his programme' and she finally loses her patience and accuses him of being unprofessional. But when she goes to their boss for help, the PI sticks to his original line that Hank should be Emily's sole support.

Pause for thought: What could Emily do next? Can you put yourself into Hank's shoes? How should the PI act in order to resolve this situation? What could have been done to prevent this situation?

The 'bottom line' can also be illustrated in the context of Belbin team roles (Chapter 9; Box 15.2).

Box 15.2
Case study 2

Peter is seriously looking for a lectureship as he has started the fourth year of his prestigious independent research fellowship. He realises that, if he could secure a new research grant to fund a postdoc, it would be a major plus for his research programme and for his CV. But, as there is only one more year to go on his fellowship, he needs to find a willing collaborator before he can even submit a grant application. The funding situation in the UK has worsened, and so Peter is considering the possibility of overseas funding.

The first opportunity presents itself at a conference in the US where he discovers that an American colleague, Ken Miller, has research interests close to his. Ken came across as self-confident, to the point of arrogance, but they got on well socially. Peter is really excited about a possible collaboration as their research area is attracting a lot of interest and the chances of funding for an international project are very good. He has already drafted a proposal and asks Emily to comment on it before emailing it to Ken.

Emily is worried because on more than one occasion, she has seen Peter get excited about a project, only to drop it when another apparently more exciting one presents itself. However, she realises that, with a suitable collaborator and postdocs, who would balance Peter's 'weaknesses', a joint project is very likely to succeed. Ken's position would be equivalent to a British PI, and there would be two postdocs on the project – one in the States, the other in Edinburgh.

Pause for thought: In terms of Belbin team roles, who would be a good collaborative partner for Peter? And what type of 'Belbin person' would be ideal as a postdoc for Peter? and for Ken?

'Community Service'

As you start to raise your profile in the scientific community, you will be asked to perform a variety of 'extra-curricular activities', e.g. reviewing papers and grants, membership of committees of learned societies, plenary lectures at conferences, public education lectures, media interviews, membership of research council committees, advisory boards of journals, etc. Many of these activities can be beneficial to your own professional and/or personal development, but not if they are taken on at the wrong time. Very few have financial rewards. In each case, you need to weigh up the cost–benefit ratio before agreeing to do anything. Below, is a range of activities that you may be asked to take on.

Giving talks

As discussed in Chapter 2, there are three main types of talk – plenary lectures at conferences, seminars at other universities/institutions and public outreach talks. Plenary lectures are very important for raising your professional profile, but unless you have enough new data to talk about, they can be really stressful. Seminars at other institutions are less stressful and, importantly, expose you and your work to a broad range of people – some of whom may be sitting on panels that are assessing your grant applications or may be refereeing your papers. Outreach activities are a necessary component of your job as a scientist, but think about what type of role you are most suited to – e.g. talks in schools or talks to adult education groups – and volunteer for those you are most likely to enjoy.

Working for journals

The three main roles you may be asked to take on for journals are - writing 'news and views' type articles, being on an advisory board or being an editor. Writing brief summaries of topical papers does not need to take a lot of time and it will force you to scan the literature on a regular basis. It will also give you practice at writing concise summaries. At busy times, it is something you can often persuade a postdoc in your group to help with. Being on a journal's advisory board gets your name in each issue of the journal, gets you a free subscription and acts as a mark of professional recognition on your CV. In most cases, it doesn't involve more than reviewing a few papers each year for the journal. Being an editor, however, can be extremely time consuming depending on the office back-up and online resources the journal has. Certainly, not a job to take on lightly, but it does act as a mark of recognition.

Writing reviews

There is no doubt that writing reviews gets you noticed by the community. Short reviews for 'Trends' or 'Current Opinions' journals shouldn't take too long to write and are read by many people. These are definitely worth doing in the early stages of your career. Longer reviews such as 'Annual Reviews' take a lot of time, but are good to do when you want to reflect on the field – it forces you to hide away and think about the science. In both cases, having a comprehensive and up-to-date electronic reference library speeds up the process significantly.

Refereeing

You will undoubtedly be asked to referee both grants and papers, and you have an unwritten obligation to do it. On the positive side, you will get to see what other people are doing/thinking about and you will also start to appreciate the difference between something that is well written and something that is not. This should help you with your own writing. It is acceptable to turn down some invitations to referee and it is better to do that if you are too busy than to return a sloppy review. You will build up a reputation as a reviewer and you want it to be a good one.

Committee work

You may be asked to sit on research council committees or committees for learned societies. Research council committees are very labour intensive but are very useful in terms of seeing how the system works. You are unlikely to be asked early on in your career but when the invite comes, accept. Committees of learned societies are less influential and are not seen to be such a mark of recognition as research council committees. However, for very little work, you can get a lot out of them. For example, you will get to work with a range of people outside of your immediate discipline, you will be exposed to discussions on science policy and you will have input into the scientific agenda of the society. In particular, you will be able to attend all of the society's scientific conferences and to meet all of the speakers. Basically, a good networking forum.

External examining

At all stages of your career, you may be asked to be an external PhD examiner. Later in your career, you may be asked to act as external examiner for an undergraduate or taught masters course. PhD examining can be quite interesting and you have some obligation to do it if you want your own students examined by other people. It will probably take about 3 days in total including travel. Examining taught courses is a different matter – it is very time consuming for little reward. Think hard before accepting.

Organising conferences

This can be fun, particularly if you have good office back-up for registrations, etc. You get to have a scientific agenda of your own and to spend time with like-minded scientists (who are often also your friends) at someone else's expense. However, start small with a 1-day conference before agreeing to organise something like a Gordon conference.

How we did it

We cover these topics in three separate workshops. For collaborations and community service, PIs with extensive experience in each area come and discuss their experiences with participants.

For the networking workshop, we devised an active listening exercise based on information freely available on the web. We asked participants to watch a video before the seminar http://www.youtube.com/watch?v=7AxNI3PhvBo (last accessed 05/3/2011). They were also asked to prepare a subject that they could comfortably talk about for a few minutes – e.g. career plans, the referee's response to their latest paper, holiday plans or hobbies. During the workshop they were asked to work in pairs, first to actively listen to a partner for about 5 minutes, and then to be listened to in turn.

Speakers had to describe their subject without providing too many details. Listeners had to listen attentively to what was being said and what was not being said – and to demonstrate that they were listening through their body language. After 3–4 minutes, the listeners had to summarise the three main issues or criteria that they had heard their interlocutors express. The speakers then had 1 minute to provide feedback as to whether the listeners had got the key points. An additional minute or so was spent on how well the listeners demonstrated active listening behaviour through information recall and body language. The speakers and listeners subsequently swapped roles and repeated the exercise. A whole group discussion to synthesise the key learning points completed this exercise.

Summary

In this chapter we look at how being an independent scientists fits in with being a member of the scientific community and the wider world – specifically, at networking, collaborating and community service. Networking increases your circle of influence and is an essential activity for any scientist. In some fields such as particle physics, collaborations are also essential but in others they are a matter of personal choice. Collaborations can lead to more effective science but can also be time consuming and ineffective – the key to avoiding the latter is to make sure that expectations of all parties are clearly understood from the start. Finally, we all have an obligation to contribute time and effort to the community that is global science. The trick is to make your contributions at a time that is right for you and your science.

Selected reading

If you only have time to read one book, make it:
Ziman, J. M. (2000). *Real Science: What It Is, and What It Means*. Cambridge: Cambridge University Press.

Other texts:
Covey, S. R. (2004). *The 7 Habits of Highly Effective People*. London: Simon & Schuster.
Covey, S. R., Merrill, A. R. & Merrill, R. R. (1994). *First Things First*. London: Simon & Schuster.

Web resources:
http://www.jobs.ac.uk/careers/ (last accessed 7/3/2011)
http://www.questcareer.com/networking_skills.htm (last accessed 7/3/2011)
http://www.businessandnetworking.co.uk/NetworkingSkillsCategory.html (last accessed 7/3/2011)

16

Designing a taught course

Good teaching practice takes years to mature, and its many elements cannot all be covered here – new lecturers should always attend the relevant introductory courses at their university. In this chapter we cover a few overriding principles of course design – getting the level right, deciding on the number of lectures to be delivered and the choice of different media. Educational theory as put forward by Ramsden is illustrated with examples.

The theory

Not everybody is both a gifted researcher and a good teacher; however, anyone can improve their teaching skills. Teaching presents complexity in its own right – it takes time and effort even to do it badly. Whereas bad teaching can put students off a subject for life, good teaching can help students improve their understanding of a subject and enable them to apply abstract principles to real problems. So, it helps to understand how students learn. In the last four decades, a lot of educational research has been carried out in this area (Ramsden, 2003). A good course should enable students to increase their knowledge of a subject to the extent that it ultimately changes their understanding of the world around them. Acceptance of this definition requires both a method to test this 'understanding' and a quantitative measure of what 'understanding' means in the context of a given course.

A summary of how students learn is shown in Box 16.1.

Let's take a look at examples of both deep and surface learning in the context of Newton's second law of motion. A student who is cramming before exams will learn the numerical formula $F = ma$. This surface approach may be enough to pass an exam, but long-term retention and the ability to apply the formula is unlikely.

In contrast, a student with a deep approach will try to figure out the meaning of the symbols, the conditions of validity and the applications. For example, s/he might calculate the impact of a car driving with a certain velocity v in a head-on collision with another car, or calculate the minimal distance needed to break safely. To do

---

Box 16.1
Approaches to learning

Students' perceptions and approaches to learning were the subject of a research project conducted through interviews in the 1970s (Säljö, 1979). Analysis of the interviews showed five distinct perceptions:

1. Learning as a quantitative increase in knowledge. Learning is acquiring information or 'knowing a lot'.
2. Learning as memorising. Learning is storing information that can be reproduced.
3. Learning as acquiring facts, skills and methods that can be retained and used as necessary.
4. Learning as making sense or abstracting meaning. Learning involves relating parts of the subject matter to each other and to the real world.
5. Learning as interpreting and understanding reality in a different way. Learning involves comprehending the world by re-interpreting knowledge.

There is a qualitative division between *surface* (focused on the external signs and learn by rote) and *deep* (focused on the meaning of the task in relation to other parts of knowledge) approaches to learning between levels three and four.

---

this, the student would have to know the meaning of acceleration ($a = dv/dt$) and momentum (mv). The role of mass m would be obvious – the heavier the car, the longer the braking path; the same reasoning applies to the speed of the car. Note the distance is hidden here, but this person is inquisitive and will figure it out. In this case, working through the applications and implications of Newton's second law will result in increased knowledge and may also change the way this student drives.

Pause for thought: Most students will use both deep and surface approaches to learning depending on the circumstances. Think back to the time when you were a student and write down a few examples of your deep and surface approaches.

Recognition that students use both deep and surface approaches to learning depending on their motivation and the nature of the task is fundamental for good course design. Deep learning implies that students fully understand the new material they are being taught and that they can relate it to already accrued knowledge. (This is the purpose of building a curriculum where the first year course is a prerequisite for the second year course.) In the sciences, a student must have factual knowledge in order to progress and it is often easy to allow students to take a surface approach, particularly in the early years of study. However, a deep approach can be encouraged by presenting the facts in relation to wider concepts, working from a well-structured knowledge base, taking care about motivational context, budgeting for learner activity persistently engaging with key ideas, and ensuring a clear and confident delivery (Toohey, 1999).

Good teaching practice promotes a quality learning experience and course design should ensure that both students and teachers get maximum benefit from their contact. So how do you do it? According to Toohey (1999) the first question every lecturer should ask is 'what is it most important for these students to know and how might they best learn it?' Straightforward as this question appears, the answer will depend on the knowledge, skill and motivation of the lecturer(s) designing the course. Most university core courses are designed in a traditional way; they are firmly rooted in the department's curriculum, their content is subject to accreditation internally (by the institution's education committee) and sometimes also externally by professional bodies. Alternative methods of course design are possible, however, and will be discussed at the end of the chapter.

The practice

Somewhat simplistically, any teaching in the sciences will have three components: the content, form of delivery and presentation. In the educational context outlined above, the difference between teaching and learning is that of efficiency and effectiveness: a lecture can be delivered brilliantly but students will not learn from it if it is not effective.

In the context of course design, a teacher therefore first needs to pose several questions to enable them to focus on the most important issues (Box 16.2; Ramsden, 2003).

	Box 16.2
	Questions to inform course design
Aims:	What do I want my students to learn?
Strategy:	How can I arrange the content to optimise learning?
Assessment:	How do I find out what students have learned?
Evaluation:	How do I find out how effective I am and how do I use this information to improve?

The first thing to establish is the place of the course within the curriculum and its 'weight'. A core course may well need more background research than an advanced option. The design of the former requires links to other courses whereas the latter may be stand-alone. There are several factors that will determine how the course is finally structured. These factors can be grouped into three categories – *external influences*, *internal decisions* and *feedback*.

External influences can include directives from organising bodies that oversee the whole degree course. For example, they may state that the course should be supported by a textbook, should be based on fundamental principles and should be relevant to industrial R&D. Internal decisions are obviously more varied and are essentially person dependent. Feedback can take the form of input from colleagues or from students who have taken similar courses.

Below, we demonstrate the design of both a core and an optional second-year undergraduate course.

Core course

Our first example is a 30-hour course on 'Fundamentals of statistical and thermal physics' taught to second-year physics students. It forms part of a core course that is common to degrees in Applied Physics, Physics with Computer Science, and Physics with Advanced Instrumentation. It builds on a first-year course 'Foundations of mechanical and thermal physics' and provides the foundation for a third-year course on 'Thermodynamics'.

Aims

The major aim of this course is to introduce students to modern concepts in physics. In this particular case the designer wants to consolidate aspects of statistical physics that she uses in her research. She is particularly keen to get the subject across effectively because it has a reputation for being both boring and difficult – despite the fact that the turbulent origins of thermodynamics and statistical mechanics were laid down by a few colourful characters, with unusual career paths. (Ludwig Boltzmann, the father of the immortal formula for entropy $S = k \ln\Omega$, committed suicide – pushed to it by envy, hostility, lack of understanding and no recognition of the importance of statistical mechanics by his contemporaries.)

In considering the course content, the designer consulted various textbooks and also thought about the key principles of statistical mechanics. After further con-sultation with colleagues, she decided on the preferred content (see Box 16.3).

So how do you go from the content in Box 16.3 to course structure? Usually, the content and structure of any 'essential physics' course is inherited from your predecessors. Unless you are familiar with the philosophy of science, the logic and hierarchy of ideas are not that obvious: you simply acquire and assimilate an overview as a student and propagate that view in subsequent research. However, it does pay off to think hard about the links and connections between the main concepts and to sketch them as a 'concept map' (see Fig. 16.1). Such a map may

---

Box 16.3
Proposed course content

Statistical methods in physics: concepts and examples, the simple random walk.
Revision of probability distributions.
Statistical description of systems of particles, microstate, statistical ensemble.
Distinguishable particles.
Entropy. Statistical definition of entropy, second and third law of thermodynamics.
Reversible and irreversible processes.
System in contact with a heat reservoir, canonical distribution (paramagnetism). Partition functions and their properties. An ideal spin – ½ solid.
BE, FD statistics. MB statistics as a classical limit.
Elementary theory of transport processes. Collision time and scattering cross-section, viscosity.

---

not be unique but it is a very effective tool for developing your own understanding of science.

> **Pause for thought:** Think about the concepts in a course that you teach or would like to teach and draw a concept map now.

Aims vs. objectives

For clarification, in an educational context aims are defined as 'general statements of educational intent' whereas objectives are 'more specific and are statements of what students are expected to learn' (Ramsden, 2003). In Box 16.4 the course aims are articulated in the first sentence. The course objectives are italicised.

Teaching strategy, assessment and evaluation

To make the content of this course approachable, the lecturer decided to deliver 20 hours of lectures and have students lead 10 hours of problem-solving seminars. Each 2-hour lecture slot was followed by a 1-hour seminar. These seminars provided intermediate links to the course material, some formative assessment and preparation for the end of course exam, i.e. the summative assessment. Evaluation was two-fold: verbal feedback after each seminar and through the end-of-course questionnaire. Students' suggestions were taken into account as the course evolved over the 10 years it was taught.

> **Pause for thought:** Think about a course that you are planning to deliver and state its aims and objectives.

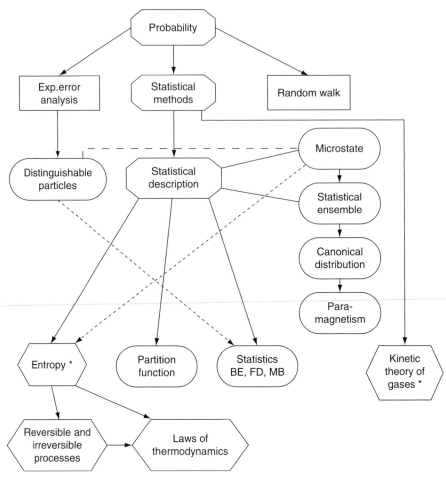

Fig. 16.1. A simplified hierarchy and interrelations in a statistical mechanics course. This is only one of a number of possible hierarchies. Solid lines – main links between concepts; dashed lines – additional relations. Starred items – independent description in terms of classical thermodynamics.

Optional course

Our second example is an optional second-year course for students studying biological sciences. Most of the students enter the degree course with an interest in animal behaviour and/or environmental science. Less than 10% of students enter the course with a stated interest in plant sciences and historically less than 20% have continued with the subject beyond the first year. The lecturer, however, is keen to ensure that as many students as possible graduate with a knowledge of the importance of plant science for solving some of the future problems of world hunger. So, let us look at the four components of course design with this in mind.

Box 16.4
Aims and objectives – 'Fundamentals of statistical and thermal physics'

1. The introductory part of the course aims at laying mathematical foundations of statistical physics. Having worked through lectures 2–7 you should be able to *define the following terms*: probability, independent events, mutually exclusive events, permutations and combinations, addition and multiplication principles, random walk, binomial distribution, normal (Gaussian) distribution, distribution function, moment of function, mean (expected) value, variance, standard deviation, dispersion, rms.
2. After studying lectures 1 and 8–30 you should be able to *define the following terms*: thermodynamics, statistical mechanics, kinetic theory, statistical mechanics of irreversible processes, system, macroscopic parameters, statistical ensemble, microstate, energy levels, spin, fundamental statistical postulate, isolated system, accessible states, density of states, macrostate, isolated system, equilibrium conditions, thermal and mechanical interaction between systems, exact differentials, entropy, absolute temperature, four laws of thermodynamics, key principle of statistical mechanics, canonical distribution, Boltzmann factor, partition function, Brownian motion, paramagnetism, Maxwell–Boltzmann, Bose–Einstein and Fermi–Dirac statistics, collision time and scattering cross-section, viscosity.
3. At the end of this course, you should be able to *calculate* probabilities of occurrence of various events, mean value, standard deviation, rms, energy and entropy of a system, partition function; *apply* the canonical distribution.

Aims

The lecturer's aim is to teach students how plant breeding programmes have been traditionally carried out and how they might be carried out in the future. Essentially – plant genetics.

Teaching strategy

The lecturer knows that the majority of the student cohort is reluctant to engage in the more molecular components of biology and is also generally weak mathematically. Since plant breeding programmes rely heavily on both skills – the course needs to be designed so that students think that the benefits of the course outweigh the negative aspects. Importantly, the course needs to cover both classical quantitative genetics and molecular genetics – most students are comfortable with one or the other, but not both.

Pause for thought: Think about the student cohort in your department. How would you persuade them, through course design, that a topic was really interesting?

In this particular case the lecturer decided that all students were generally interested in: people and future global problems (e.g. climate change), food (or lack of it), money (or lack of it) and sex. So, how can you design a course that aims to teach plant genetics around these four components? Well, genetics and sex are essentially inter-related, so that part is easy. Making students realise that plants have an enormous impact on people is not as easy as it sounds, so the lecturer called the course 'Plants and people' on the grounds that it should at least make the students look at the content to see what they might be missing. The next design step incorporated interest in food and money by deciding that each lecture should be based around a particular food crop, and that a component of each lecture should include the global economic impact of not being able to harvest enough of that crop in any particular year. Incorporating an historical component of domestication in each case would also introduce a people component to what is essentially quantitative genetics.

In this case, the lecturer would have to learn some economics and some history to deliver the course – a case of expanding your own knowledge through teaching.

Assessment and evaluation

So, how can the lecturer find out whether students have actually learnt plant genetics as opposed to going away with a number of anecdotes about the food on their dinner table, some knowledge of global wheat markets and knowledge of where and when crops were first domesticated? Assessment methods are key here.

> **Pause for thought:** What type of assessment would you set?

In our view this is a case of setting the problem in terms of the more general interest components of the course but requiring the answer to be based on plant genetics. For example:

If China started to import rice on a regular basis, it could afford to pay more than other importing Asian countries such as the Philippines, and this price increase could have a devastating effect on the Asian economy. Given the predicted effects of climate change, how would you generate new rice varieties that would enable Asia to grow and harvest more rice, on less land and with less water.

Such an assessment could be set as an essay question or could be part of a small workshop where different groups came up with different solutions.

Alternative methods of course design

Thus far, we have assumed that course design and delivery will be done by a single individual. This does not need to be the case, in that courses can be designed by teams. In the UK, this approach was pioneered and perfected by the Open University (OU). The OU courses are some of the best available and their mode of delivery differs significantly from that of typical university courses. From the very beginning, the course materials were produced by teams of people. Typically, there is an author of a particular topic and a second reader who feeds comments to the author. If time permits, other team members read the material critically and also feed comments back to the author. There may also be another person writing exercises on the same topic, and there are also tutors willing to beta-test material before it is accepted for printing. A designated team leader coordinates the work until it is finally fit for launching.

There are several benefits of this way of working. Firstly, it pulls together different areas of expertise that may not be available to a single individual. Secondly, through discussions and beta-testing, most of the ambiguities are clarified before the course material reaches the student and often more elegant explanations are found. However, there are also drawbacks. For example, if someone does not deliver on time, it can set back a part of the course, and often the initial vision of one person is diluted. In our view the benefits of this way of working outweigh the disadvantages.

Pause for thought: Look at an example of written course material that tries to explain something. Is it self-explanatory? How would you change it to make it clearer?

How we did it

Participants were split into two teams prior to the seminar. Both teams were asked to design a core first-year course of 40 lectures. In the seminar the two teams each presented the content and rationale of their course. Discussions then followed about the strengths and weakness of each proposal.

Summary

In this chapter we have briefly described the process of course design and illustrated it by examples of two very different courses. One is a second-year core statistical and thermal physics course, the other is an optional plant science course for second-year biology undergraduates. Both courses aim to motivate and engage students with the side effect of enriching the lecturers' own knowledge through teaching. The essential role of assessment and evaluation is acknowledged. When you are new to course design, the amount of effort needed can be demotivating. However, you will soon realise that the process is not smooth; the course will constantly evolve until it is deemed 'right' for its scheduled lifetime.

Selected reading

If you only have time to read one book, make it:
Ramsden, P. (2003). *Learning to Teach in Higher Education*. London: Routledge-Falmer.

Other texts:
Knight, P. (2002). *Being a Teacher in Higher Education*. Buckingham: Society for Research into Higher Education and Open University Press.
Säljö, R. (1979). Learning about learning. *Higher Ed.*, 8(4), 443–51.
Toohey, S. (1999). *Designing Courses for Higher Education*. Buckingham: Society for Research into Higher Education and Open University Press.

17

Giving a good lecture

When designing and delivering a good lecture, the same principles hold as for delivering a research seminar (see Chapter 2). However, you also need to consider how to enable effective learning. In this chapter we provide practical advice on how to maximise student learning from lectures.

The theory

Everybody recognises good teaching intuitively, and its qualifying characteristics have been confirmed by educational research. In words attributed to Richard Feynman, one of the greatest physicists and a gifted teacher: 'There are two types of scientists – one says look how clever I am and the other says look how simple it is' (Leighton & Feynman, 1992). Obviously, the latter has a much better chance of engaging students. Such an attitude reflects the teacher's enthusiasm for the subject, and also a willingness to work hard to make the subject matter accessible. But how does this teaching attitude translate into effective lecturing?

Since the Middle Ages, lecturing has been one of the principal methods of 'knowledge transfer' to students at universities (Bligh, 1998). Its prominent place in mathematical, physical and life sciences is implicitly acknowledged and it is seen as the core method, to be complemented by tutorials, classes, laboratories and fieldwork where appropriate. But what is the purpose of a lecture? Lectures are most effective when they are used to deliver new information, provide an overview of a subject or to weave together several ideas from different sources. A more mundane use is to explain key concepts in detail or to guide students through detailed calculations.

The practice

If you are asked to step in for somebody to deliver one or two lectures, you will most likely be given the lecture notes and other material presented by the usual lecturer.

What next? Your first thoughts will probably be about the content – how familiar are you with the subject and scope, can you follow the lecture outline and arguments, are there any associated materials that you need to be familiar with? But, it is also important to consider how this lecture fits into the bigger picture.

Intellectual context

Even an occasional lecture should be set in context – you need to consider the students, your colleagues and department, and the overall course structure. The minimum information you need about the students is some understanding of how they learn and of their educational background. The number of students and year of study is also important. Your colleagues and department will determine the ethos – is lecturing seen as a chore by the majority, or is it seen as an essential means of communication with students? The place of your lecture in the overall course structure is also clearly important – is this lecture in the middle of the course, or is it a summary of the course? Is this an optional or a core course? Are there any associated classes and practicals? Has the course got clear objectives, and how do they link to this particular lecture?

> **Pause for thought:** Think of the context in which your lecture will be delivered. If you do not have to deliver one in the near future, think about the context of somebody else's lecture.

If a lecture is part of a course, the first thing to do is find out how it fits in with the rest of the course, and discover what the overall course objectives are. It is often the case for a new or a visiting lecturer that there is not much freedom to change the scope of a lecture first time around. However, it is usually possible to change the way the subject is presented or to introduce some new elements so that you feel more comfortable with its delivery. You may also be able to refine the objectives.

The student cohort

The characteristics of the student cohort are the next factor for careful consideration. The purpose of their studies will influence the way they will perceive and react to your message. First-year undergraduate students typically look to a lecturer as a figure of authority, whereas final-year undergraduates or first-year graduate students will be more realistic and forgiving. If this is a first-year undergraduate lecture, then the students' educational background and the course entry criteria should influence the lecture content. The students may be young, immediately post-school, highly motivated and ambitious. Alternatively, you may find out that, because of a wide access policy for the course, you will have to address students from a mixed educational background. In both cases, students may need some help with the transition to university teaching, both in the technical and psychological sense.

> **Pause for thought:** Think about and write down what you know about the students that you have been asked to teach.

Student motivation

Initiating and maintaining student motivation is a big issue, and most lecturers struggle with it at times. In our experience, during a 3-year course, motivation often flags during the second year. The excitement of getting into university is gone, there is no end in sight and often the subjects taught cover the middle ground between introductory and really interesting advanced material. There is no panacea – but carefully laid-out lectures, with clear objectives and support material help. It also helps to be aware of how students' attention spans vary during a lecture. Lectures are typically 50–55 minutes long. Within that timeframe, student attention spans decline after the first 20 minutes, and for the rest of the lecture they will not pay attention for any more than 10 minutes at a time. So, it makes sense to introduce and explain the most difficult concept within the first 20 minutes and then to have a short pause. This pause can be used to obtain student feedback (e.g. ask 'is everything clear?') or to allow them to think back and attend to their notes. Some lecturers tell a joke or a little story, or simply open a window to signal a mini-break.

Practical context

The actual setting for a lecture is important and can make a big difference to how you feel. While you may not have a choice, it does help to be aware of the advantages and disadvantages of the physical parameters of any particular room. After all, it is more difficult to feel comfortable, while lecturing to 200 students in a steep amphitheatre than to 20–30 in a more intimate setting. The setting also influences the verbal and non-verbal feedback you can get from students during the lecture – how easy is it to make eye contact with somebody 20 metres away?

> **Pause for thought:** Write down a checklist of things to do and to take with you to your lecture.

Lecture preparation

Going through the pre-lecture checklist in Box 17.1 should result in a clear picture of 'what' and 'where'. You can now start preparing the lecture by thinking it

---

Box 17.1
Example of a pre-lecture checklist

- Locate the lecture theatre and the key/card to open it.
- Check what equipment is available and learn to operate the data projector/video and equipment/lights.
- Make sure there is a set of working pens/chalk to write on the whiteboard/blackboard.
- Get the names and location of people who can help with technical problems.
- Find out the seating plan and the number of students expected to attend your lecture.
- Check the lecture schedule – will you be the first to lecture in the room on the day or will you have to wait until the previous lecture has finished?
- If using your own laptop, think about what you will do if it fails – is there a spare or dedicated 'teaching' computer? Have your lecture backed up as a PDF on a USB pen just in case all else fails.
- Print out a copy of your lecture notes to take with you.
- Bring a sufficient number of student handouts with you.

---

through. It is normal to start with the objectives (Bligh, 1998), but you may find it easier to start thinking about the key concepts to be introduced or discussed. The essential thing to remember is that any good talk has three stages – introduction, subject matter and conclusion. This is sometimes referred to as the 'rule of three'. In terms of undergraduate lectures you can think of it as follows – tell them what you are going to talk about, tell them about it, and then tell them what you have told them. At each stage, you need to use clear slides with simple illustrations and minimal text. Estimates for preparing a 1-hour lecture first time around vary from 8 hours to 3 full working days and more. We think it is time well spent, especially if you need to give the same lecture again. After all, it will clarify your own thinking and consolidate your knowledge.

> **Pause for thought:** Look at a copy of your lecture notes. Can you identify objectives and key concepts? Analyse its structure – does it comply with the 'rule of three'? In a suitably private setting, rehearse your presentation out loud, and estimate carefully the amount of material you should prepare for the time available.

Effective teaching

It is common for many good lecturers first to come to grips with their lecturing technique before thinking more in depth about the effects of their lecturing. So, you will probably not worry how effective your lecturing is at first, but simply be relieved that it more or less went according to plan and that you survived the experience. But, even at this stage it is worthwhile being aware of what students are likely to get out of lectures, because teaching is not the same as learning. According to Ramsden there are six key principles of effective teaching (Ramsden, 2003). We outline these below.

Interest and explanation

In terms of the student experience, the first principle is that of interest and explanation. Students state that the essential characteristic of a good lecturer is the ability to explain things clearly, and that they are kept motivated by a lecturer's enthusiasm and interest in the subject taught. Despite the hard work needed to master the skill, if you can successfully make a subject interesting for students, then they will find it much easier to learn. While we do not have a magic formula for enthusiasm – in our own teaching we use examples from our own research that are of obviously interesting to us – the art of clear explanation can be learned (see Box 17.2).

Box 17.2
Making a difficult concept more accessible

A typical problem in physical and mathematical sciences is the derivation of a very complex equation. There are two ways of doing it. The first is to set the scene and build the equation from scratch. This route may follow historical developments, and is often compared to the path up to a mountain summit – the summit being the equation. The other way is to introduce the equation and to discuss its meaning, consequences, limitations and applications. This is the path from the summit on to the plain.

Each way has its merits, and it is desirable to mix the two approaches, especially in a longer lecture course. If the first approach is chosen, it helps to show the summit from the base camp. During the arduous climb, the glimpses of the summit and the signposts keep the students going, as there may be many obstacles and clouds obscuring the final goal. Having got there and having had some rest, then one can admire the view. In the second approach it helps to introduce the most important parts of the equation first rather than the whole equation. In analogy with being at the summit – what are the most important factors to consider before starting your descent? Is it the altitude, the distance or the angle of the incline? You can then introduce less important quantities, coefficients, etc. and finally show the equation in its full glory.

In both cases it is helpful to visualise the equation, especially while discussing limiting cases. There are packages such as *Mathematica*, various resources on the web and maybe departmental resources to help with this – or you may be a skilled programmer yourself.

Pause for thought: Think how you would introduce a key concept or equation with a 'way up to' and a 'way down from' the summit approach.

Concern and respect for students and their learning

This principle highlights the fact that the attitude of a good teacher is like that of a good gardener – nurturing the plants and enjoying seeing them grow. It means caring about the effect of your teaching and having the will to make learning possible, and your subject accessible. Good teachers will never make any subject, no matter how difficult, look forbidding. They will also invite questions and make time for students. The antithesis of such a teacher is an arrogant, haughty 'professor' who suffers the necessity of teaching, but wishes to be somewhere else.

Appropriate assessment and feedback

Setting meaningful assignments and giving helpful feedback are critical factors in helping students to learn. Clearly, there are issues with time and resources available – the level of feedback you can give to six students is not possible for 200 students. The key is to set assessments in a way that allows you to provide as much feedback as possible to every student.

Clear goals and intellectual challenge

This principle centres on a tension between the discipline imposed by a teacher and the freedom enjoyed by a student to discover new horizons on his or her own. It is a gradual but non-linear process and its control should rest with both parties. It is more in line with tutoring than lecturing as it often involves pastoral care and 'stretching' the student beyond their perceived limit. A lecturer has to be able to explain clearly what students are required to learn in order to achieve understanding at this stage and to set the scene for what follows later.

Independence, control and engagement

This principle crowns good teaching as it engages students with the subject matter through judicious choice of learning tasks. The ultimate goal is to free students from over-dependence on teachers. We examine methods to achieve this goal in Chapter 18.

Learning from students

What – if anything – can we learn from students? As a final comment in this chapter, we provide a quote from Richard Feynman (Leighton & Feynman, 1992).

If you are teaching a class, you can think about the elementary things that you know very well. These things are kind of fun and delightful. It doesn't do any harm to think them over again. Is there a better way to present them? The elementary things are *easy* to think about; if you can't think of a new thought, no harm done; what you thought about it before is good enough for the class. If you *do* think of something new, you're rather pleased that you have a new way of looking at it.

The questions of the students are often the source of new research. They often ask profound questions that I've thought about at times and then given up on, so to speak, for a while. It wouldn't do me any harm to think about them again and see if I can go any further now. The students may not be able to see the thing I want to answer, or the subtleties I want to think about, but they *remind* me of a problem by asking questions in the neighborhood of that problem. It's not so easy to remind *yourself* of these things.

So I find that teaching and the students keep life going, and I would *never* accept any position in which somebody has invented a happy situation for me where I don't have to teach. Never.

How we did it

This seminar followed on from the designing a course seminar (Chapter 16). Each of the design teams nominated an individual to deliver a lecture from the course, but all team members contributed to the preparation. Each team thus presented a 40-minute lecture that was assessed by the other team for effectiveness.

Summary

In this chapter we have provided practical tips for lecture preparation and delivery that are based on the 'rule of three' – introduction, subject matter and conclusion. In this context you should remember that less is more – don't overload students with information, that 'pictures are worth a thousand words', and that the considered use of different presentation media will help keep students engaged. In the end, however, it will be your enthusiasm and commitment to teaching that will determine how successful you are.

Selected reading

If you only have time to read one book, make it:
Bligh, D. A. (1998). *What's the Use of Lectures?* Exeter: Intellect.

Other texts:
Leighton, R., & Feynman R. P. (1992). *Surely You're Joking, Mr. Feynman! Adventures of a Curious Character.* London: Vintage.
Ramsden, P. (2003). *Learning to Teach in Higher Education.* London: Routledge-Falmer.

Web resources:
Higher Education Academy (HEA) Subject Centres for:
Bioscience – www.bioscience.heacademy.ac.uk/resources
Engineering – http://www.engsc.ac.uk/er/snas/index.asp
Information and Computer Science – http://www.ics.heacademy.ac.uk/
Materials – http://www.materials.ac.uk/
Mathematics and Statistics – http://www.mathstore.ac.uk/
Physical Sciences – http://www.heacademy.ac.uk/physsci/

18

Beyond lecturing

In the sciences, lectures are complemented by classes, laboratory or fieldwork, seminars and different types of projects. All of these methods are meant to foster increasingly independent learning as well as provide a sound basis for being a proficient scientist. All of the methods have their merits and problems, but at the very least they should engage the student and have meaning and relevance to material given in lectures. In this chapter, we discuss small group teaching methods, inspect distance learning and give a brief overview of the possibilities offered by interactive teaching. The pedagogical principles expounded in the previous two chapters underpin this discussion. We finish by finding out what students have to say about teaching and learning.

The theory

Giving lectures is the mainstay of an academic profession – in the UK the most junior long-term or permanent position is 'lecturer'. In Chapters 16 and 17 we discussed how to design, prepare and deliver lectures whereas in this chapter we turn our attention to other methods that can bring about effective learning. It is important to remember that a sound teaching strategy is more important than the delivery method (Ramsden, 2003). The assumption that a method, particularly a technologically advanced one, is the key to effective learning is false.

The practice

Small group teaching

Somewhat simplistically there is a basic difference between lecturing and teaching a small class: in the former the lecturer speaks *ex cathedra* and there is often a physical distance between the lecturer and the audience. If, in addition, there are 200 students or so, opportunities for interaction (asking questions, for example) are rather limited. Although it is not impossible, attempting to interact with a large audience is not for the shy as controlling the situation takes some experience. In class-based teaching, however, the lecturer is more approachable and even the room setting can encourage closer teacher–student interaction, and peer learning through work in small groups.

Here we define a small group as being between 10 and 15 people. This size allows for interaction between students in pairs/threes/fours. There are a few grounds rules to remember:

- Know your stuff inside out, otherwise you risk damaging your credibility.
- Define both the scope and the limitations of your class.
- Be clear in setting aims and objectives.
- Get the students to talk, listen to them, allow for peer interaction and give them feedback.

A useful pair of overriding questions to ask yourself, to inform your preparations, is: 'What is it that I want my students to be better able to do (at the end of this group session)?' and, 'How will I ensure they have the opportunity to rehearse this during the time we spend together?'

Pause for thought: Look at a lecture that you have given recently or at one given by a colleague from your department. What type of questions/problems could you set for a small class?

Work in small groups promotes deep learning, and its value is well documented in the social sciences and humanities (Bogaard *et al.*, 2005). In terms of the best method of small group teaching, however, students and tutors have different views. In one study, 67% of student respondents saw tutor-led discussion as the most effective teaching method, whereas only 35% of lecturers concurred (Bogaard *et al.*, 2005). In Box 18.1 we quote student and lecturer perceptions of what such a session should entail.

How can we apply these findings to teaching small groups in the sciences?

Small group teaching should encourage active learning through peer learning, problem-solving and feedback responses to the teacher. As with all other face-to-face types of teaching, the success depends on the attitude and empathy of the teacher, and also in this case on their ability to facilitate rather than command. There are known problems with learning in this set-up – the most common ones are that the students do not prepare for the class, do not want to talk or want to be given a solution rather than discuss the question. There is therefore always a danger that the teaching will degenerate into lecture-style delivery – but try hard to resist. One way to ensure that students contribute is to ask a slightly controversial question and

---

Box 18.1

The teacher's role in a small group setting in the humanities and social sciences

Lecturers almost unanimously responded that the tutor should be a 'facilitator' who should 'provide the session with structure', 'get students to think and talk', 'ask and answer questions', 'suggest other directions for the discussion' and, if necessary, 'make corrections'. The tutor 'should not turn the session into another lecture, as this would be both inefficient and a wasted opportunity for the students to develop their own ideas'. Only three tutors referred to a need to change their role for different academic levels.

Students thought that the tutor should be a 'leader', 'guide' or 'initiator', and assumed that the tutor should be the 'source of knowledge' and/or 'authority' within the session. Students considered it essential that the tutor provide them with information, given the perceived unreliability of the other students. There was a strong emphasis on academic progression, with the role of the tutor changing from straightforwardly didactic to detached and observational.

---

then ask each participant in turn for their opinion of the answer. This approach acts as an icebreaker and class discussion often flows more freely afterwards.

In preparing for small group teaching, your best approach is to talk to more experienced colleagues and to participate in a seminar on teaching small groups if one is organised by your institution. There is also some advice in the educational literature. Here we describe one approach to encourage peer learning that has worked for us.

The problem-solving seminars were linked to the 'Fundamentals of statistical and thermal physics' course described in Chapter 16. Every 2 hours of lectures were accompanied by a problem sheet thematically linked to the most important topics. In other words the lectures explained the principles, and the classes dealt with practical applications and further discussion of topics. Rather than leaving each student to attempt the whole list of problems, at the beginning of the course students were asked to organise themselves into groups of three or four people. Every group then picked one or two problems, undertook to solve them before the seminar and to present the problem and solution to the whole class. Students from each group would meet in their own time to work on their assigned problems. The lecturer was available prior to the seminar to give guidance. In the seminar a general discussion followed the group presentations. At the end of the seminar, solutions were swapped between the groups. There was no marking involved because the seminars did not form part of a continuous assessment process. Another approach is called for if the classes or seminars are to be assessed. One possible way is to get assignments in before the class, mark them and make a note of common problems to be discussed during the seminar.

It could be argued that the example above was an imperfect solution because not every student was equally active in the problem-solving exercise. Initially, there was a problem with some students not 'remembering' what they had committed themselves to. This was quickly solved by asking students to put their names against a given problem on the list. In the second year of running this system there was one class for which nobody came prepared. The lecturer calmly recalled the agreement that students were responsible for this class and then left the room without dealing with the list of problems. After that, the students always prepared for the class and the following years' cohorts also respected the agreement.

> **Pause for thought:** Think of a key topic in one of your lectures. How can you illustrate it with an exercise? Can you think of a way to link it to other domains of a student's knowledge relevant to this topic?

Distance learning

Distance learning means that students can study in their own time, anywhere, using course material that is either printed or delivered online. It poses additional educational challenges as the course material has to be self-explanatory to a higher degree than necessary with a live lecturer available for additional explanations. It is aimed primarily at part-time students and it relies on their motivation and an ability to juggle various commitments – in return it offers flexibility.

The Open University (OU) (www.open.ac.uk) is a world-class leader in distance learning. It was established in 1969 with the mission 'to be open to people, places, methods and ideas' and ever since then it has pioneered new teaching methods using a range of communication technologies. The OU is now so firmly established in the UK educational landscape that it is difficult to believe what a revolutionary concept it was in the 1970s. It has become famous for the quality of its educational programmes that are produced by a dedicated BBC unit and were broadcast on national TV, albeit at ungodly hours. Degree courses are offered all over the UK and are supported by locally hired part-time tutors. Students get printed materials and online access, and can sign up for optional scheduled tutorials. They also have access to a tutor via email or phone. Tutors are trained in the delivery of their course, are given guidelines and model solutions, and are periodically evaluated by staff tutors. Evaluators review the feedback that tutors provide on students' assignments and also occasionally attend tutorials. Students travel to nominated local examination centres to sit their exams. In combination, these methods achieve consistent teaching standards, whilst simultaneously allowing tutors to adopt different teaching strategies. Whereas initially the majority of students were mature and part-time – specifically those who missed out on education earlier in their lives – there is now a constituency of students starting to study with the OU directly after school.

Interactive teaching

The invention and subsequent widespread use of computers in the second half of the last century has had a lasting impact on higher education. One of the first attempts to encourage interactive learning was the use of computer assisted learning (CAL). While the principle is excellent, the method has not been widely adopted, partly because the development of CAL for any particular course is costly in terms of staff time, but also because really good packages can take many years to develop. A better formula is to mix traditional teaching with resources available on the internet – of which there are too many to list. Many institutions give free access to their educational resources and increasingly more textbooks have a dedicated website where students can view more complex material. This is especially important for biosciences, as exemplified by a website for biochemistry (www.whfreeman.com/biochem6), which allows further studies of conceptual and structural insights, the viewing of living figures and the use of animated techniques (Berg *et al.*, 2007).

One of the most effective ways of encouraging interactive learning involves the use of 'clickers' in a traditional lecture context. (Clickers can also be referred to as audience paced feedback, classroom communication systems, personal response systems, electronic voting systems, student response systems, audience response systems, voting-machines and zappers). Clicker use is very popular in the US but seems to be used infrequently in the UK. The clicker system consists of software installed on the computer used for presentation, a wireless 'hub' and hand-held personal eggs with several buttons. During a lecture, multiple choice questions can be asked and students vote for their answer by pressing the appropriate button. Appropriately framed questions allow the lecturer to check how well students have understood and retained information given either in the same lecture or in the preceding one (see examples in Box 18.2). Clicker use also serves to engage student attention, and can provide a welcome break from passive listening.

Box 18.2

Example of multiple choice question for use with clickers

Thermosetting polymers have the following structure:

1. Long flexible chains – weak interactions via long-range van der Waals or electrostatic forces.
2. Physically entangled or lightly cross-linked (vulcanised).
3. Strongly cross-linked through chemical reactions – a rigid 3D network, covalent bonds.
4. Don't know.

Clickers can also be used to get feedback for the revision of course topics, and to encourage student participation in general discussion. As they are numbered, they can also serve administrative purposes, for example in continuous assessment. A recent review details the use of clickers in several disciplines with the main emphasis on teaching chemistry (MacArthur & Jones, 2008). It is estimated that there were about 8 million clickers in use across US campuses in 2008 and they were reportedly popular with students.

Pause for thought: Imagine that you are giving a lecture and you want to check that the students have understood a key point. Write down a question that could be used with clickers.

Students' views on teaching and learning approaches

The UK Higher Education Academy (HEA) recently carried out a review of the student learning experience in physics (Edmunds, 2008) and chemistry (Gagan, 2008). It was conducted by means of questionnaires and subsequent selective interviews. The survey acquired a statistically significant number of respondents – some 700 students and 289 staff for physics, and 332 students and 237 staff for chemistry. Both students and staff agreed on the effectiveness of learning in small groups and through individual project work. Problem-based learning was also seen as an essential part of the student 'toolkit', conducted in the context of tutorials, workshops and laboratory programmes. e-Learning, however, received a very mixed reception – chemistry students found e-learning the least enjoyable and effective teaching method. They did, however, appreciate online access to external resources and to university libraries.

How we did it

We organised a day-long workshop on 'Teaching practices in the sciences' in colla-boration with the Oxford Learning Institute (OLI), where participants could learn about teaching methods other than lecturing. Work in small groups was facilitated by experienced senior colleagues from OLI and several departments. For interactive teaching, one of us used clickers in lectures to first-year and third-year students. Both cohorts found them fun to use and helpful in understanding lecture topics.

Summary

In this chapter we have discussed the merits of small group teaching, distance and interactive learning and the use of using electronic technology. There is no doubt that, with the wealth of information available on the internet, opportunities for self-learning are much greater than ever. However, learning is still enhanced through teacher–student interactions and, if we view teaching as a conversation between teacher and student, then it is obvious that dialogue flows more easily in small groups. The key point to remember is that, in small groups, the teacher should facilitate and not command.

Selected reading

If you only have time to read one book, make it:
Ramsden, P. (2003). *Learning to Teach in Higher Education*. London: Routledge-Falmer.

Other texts:
Berg, J. M., Tymoczko, J. J. & Stryer, L. (2007). *Biochemistry*. New York: W. H. Freeman.
Bogaard, A., Carey, S. C., Dodd, G., Repath, I. D. & Whitaker, R. (2005). Small group teaching: perceptions and problems. *Politics*, 25, 116–025.
Edmunds, M. (2008). *Review of the Student Experience in Physics*. P. S. Centre, Hull: Higher Education Academy.
Gagan, M. (2008). *Review of the Student Experience in Chemistry*. P. S. Centre, Hull: The Higher Education Academy.
MacArthur, J. R. & Jones, L. L. (2008). A review of literature reports of clickers applicable to college chemistry classrooms. *Chemistry Education Research and Practice*, 9, 187–95.

19

Mentoring

As you become more senior, you may be asked to become a mentor to somebody recently appointed in your department. In this chapter we investigate what makes a good mentor, and assess the benefits for both sides of the mentor–mentee relationship.

The theory

The etymology of the word *mentor* goes back to ancient Greek. Mentor looked after young Telemachus, son of Odysseus, when the latter was away fighting the Trojan war. According to mythology, the goddess Athena impersonated Mentor when the going got tough. Over the centuries a mentor came to signify somebody experienced who takes a kind interest in a less experienced person, offering guidance and support.

In the web of learning, where does mentoring belong? There are four distinct ways of helping others to learn – by being a teacher, tutor, coach or mentor. The teacher–pupil relationship is distant and is dominated by the teacher providing explicit information to the pupil. In the tutor–student relationship, the student goes beyond the given facts and hones his or her understanding through discussion, whereas in coaching, knowledge is transferred through demonstration and feedback to the learner. Mentoring differs from all of the other forms in that learning is less tangible – intuitive knowledge and wisdom are transferred in an environment of encouragement and stimulation. Wisdom here means the ability to apply accumulated knowledge and skills to a new situation (Clutterbuck, 2004).

A good mentoring programme will have four clearly identifiable phases – beginning, middle, end and informal follow-up (Clutterbuck, 2004). In the first phase, the main role of the mentor is to establish a rapport with the mentee and to develop a sense of purpose to the relationship. In the second stage, the mentor starts to share their own experiences, and offers encouragement and constructive criticism to the mentee. This phase is often mutually beneficial in that the mentor can learn from the mentee. In the third, winding down phase, the mentor increasingly steps back and helps the mentee find new sources of learning. In the final phase, if there is mutual trust, the relationship can continue informally for many years with infrequent meetings and the mentor serving as a sounding board for new ideas.

The practice

The mentoring relationship

The first thing to consider when you are asked to act as a mentor is whether you are ready for this role. What is required of a mentor, and who needs to be mentored? Most organisations, including higher education institutions, have mentoring schemes for newly appointed staff. The common aim of all successful mentoring programmes is to get new staff settled in so that they more rapidly achieve their full potential. Would you be able to spend time on this activity? Is your knowledge of your department/institution/professional field such that you can guide others?

As people have different ideas about what mentoring entails, expectations on both sides of the mentor–mentee relationship vary widely. In an ideal world the mentor and mentee would be carefully matched, and every institution would have an extensive training programme for mentors. In reality, in a small department there may be just one or two individuals who are willing and able to spend time helping somebody adjust to a new work environment.

So what are your views of a mentor–mentee relationship?

Pause for thought: What are the characteristics of a good mentor (suggestions in Box 19.1)?

> Box 19.1
> **In your opinion, a mentor is ...**
>
> - A friend or companion.
> - Your PI (line manager).
> - A role model for good practice.
> - A coach to help you learn.
> - An expert and source of professional advice.
> - An assessor.
> - A counsellor.
>
> Add any other characteristics that you think are important.

Pause for thought: If you have been mentored in the past, what were your expectations of your mentor? If you are not able to identify any mentors either in the past or now, how do you think a mentee should behave (Box 19.2)?

Box 19.2
In your opinion, a mentee should ...

- Seek help from the mentor.
- Treat the mentor as a source of technical information.
- Ask for meetings when they have problems.
- Follow the mentor's advice indiscriminately.
- Think of how the relationship could benefit both the mentee and the mentor.
- Demand complete confidentiality even if a problem needs outside intervention.
- Decide when to terminate the mentoring relationship.

Pause for thought: What features do you think define a good mentoring relationship (Box 19.3)?

Box 19.3
In your opinion, the mentoring relationship should ...

- Be formal.
- Be informal.
- Be based around a schedule of regular meetings.
- Have a clear beginning, middle and end to the programme.
- Be totally confidential.
- Be partially confidential in that the mentor should notify appropriate people if the mentee has a serious problem.
- Be a route to report problems to the relevant head of department.

The mentoring process

Mentoring in academia tends to be somewhat informal, whereas in industry or governmental institutions it is formal, with training for both mentors and mentees. Formal mentoring programmes may even involve signing a contract where both sides agree how often and for how long they should meet, what the aims of the relationship are and when the official mentoring relationship is expected to end. Whilst formal contracts are unlikely to be encountered in academia, there are a number of important issues that must be agreed on. The first and most important is confidentiality – the mentee should feel that they can discuss problems without their line manager being informed. Another issue is mutual respect for the time being spent – agreed meetings should not be cancelled or rushed without a serious reason for doing so. There should also be an agreement about the length of the mentoring programme – in academia this can be linked to the probation period, so it may run from 6 months to 3 years.

Regardless of how long a mentoring programme lasts, it is important to use the time well. It is therefore helpful to agree topics for discussion ahead of time so that meetings can be clearly structured. In Box 19.4 we list examples of discussion topics.

Box 19.4
Examples of topics for discussion

This list is not meant to be prescriptive, merely illustrative.
Research: Research plans.
Teaching: Preparation and observation of lectures, teaching methods – classes, laboratory-based teaching and tutorials, supervision of graduate students.
Assessment: Assessment methods, design of examination papers, marking standards, question setting, examining of research degrees.
Administration: Departmental and university procedures and conventions.
Career development: Balance of duties, professional development.

The mentoring method

Differences between mentoring and sponsoring

The main difference between mentoring and sponsoring is that the former emphasises personal accountability and empowerment of the mentee, whereas the latter makes effective use of the power and influence of the sponsor (Clutterbuck, 2004). For example, a sponsor will intervene on behalf of a protégé in the case of conflict with others. Sponsoring is prevalent in the US, and is quite hierarchical, with the sponsor being older and more senior than the protégé. A successful sponsorship is built on reciprocal loyalty, and can be quite stifling if the sponsor is not ready to let the protégé make his or her own mark. Overall, sponsoring is a one-way learning process whereas mentoring is a two-way learning process, where the mentor helps the mentee to decide which plans to follow.

Differences between mentoring and coaching

Mentoring and coaching as ways of helping others to develop are frequently confused. Some tools are shared between the two approaches, but coaching is focused on the task-in-hand – on developing skills to improve job performance. It is essentially akin to coaching in sport, but with the emphasis on mental rather than physical skills. In an academic context, coaching forms part of the supervisor–PhD relationship where the supervisor is expected to give feedback to the student. Discussions are likely to be centred on technical aspects of research and on the goal of successful thesis delivery. In contrast, mentoring goes beyond short-term goals, feedback goes both ways and the process requires reflection by the mentee.

The benefits of mentoring

What benefits can you get from mentoring, and what benefits will your mentee get? And what about the cost–benefit balance?

Most mentors gain personal satisfaction from helping somebody to develop and become independent. For an academic mentor there is also intellectual satisfaction in helping somebody to work through a problem. There is also the benefit of seeing things from a different perspective – a perspective the mentor might find useful when facing their own challenges. For the mentee, the benefits include developing professional networks, gaining greater self-confidence, and learning how to make sense of the work environment by merging information from different sources. Often, one of the most appreciated benefits for the mentee is the knowledge that somebody believes in them.

There are also potential downsides for both mentor and mentee. For example, an overeager mentor who steers a mentee too closely is more of a curse than a blessing; if there is a conflict between the mentor and the line manager then the mentee may suffer; and if the mentor only gives advice rather than letting the mentee consider alternatives, the advice may turn out to be wrong. For the mentor, the greatest disadvantage is having an over-demanding mentee – a time-sink; and a potential pitfall for both sides is a breach of confidentiality. To a large extent, however, these disadvantages can be avoided if there is a well set-up mentoring programme in the department, and the expectations of both sides are made clear from the outset. With this proviso, the benefits of mentoring will outweigh the costs, and both the mentor and mentee will derive satisfaction from the relationship.

How we did it

We have not held a specific workshop on mentoring. However, during the course of other workshops, participants were asked to discuss how their mentor (if they had one) had advised them to tackle certain situations.

Summary

Mentoring helps people reach their full potential by providing advice and guidance, and is especially important for newly appointed staff. In most cases the process is informal, but it is important to establish the rules of engagement from the beginning (e.g. confidentiality, frequency of meetings, etc.). Whilst there are no prerequisites for being a good mentor, it is not a task that should be taken on by a direct supervisor. Essentially, mentees require an experienced person who is interested in their personal and professional development and who is willing to give time to them.

Selected reading

If you only have time to read one book, make it:
Clutterbuck, D. (2004). *Everyone Needs a Mentor: Fostering Talent in Your Organisation.* London: Chartered Institute of Personnel and Development.

Other texts:
Alred, G., Garvey, B. & Smith, R. (1998). *The Mentoring Pocketbook.* Alresford: Management Pocketbooks.

Part III

Managing your career

'What lies behind us and what lies before us are tiny matters compared to what lies within us.'

Ralph Waldo Emerson

In the first two parts of this book we have discussed how to become an independent scientist who can run a well-funded research group, publish papers of international quality, teach and train the next generation of scientists and contribute to the scientific community on local, national and international levels. In the final part we look at more long-term issues. Chapter 20 embraces Parts I and II by discussing ways to manage stress – an essential skill to acquire in any profession. Chapter 21 then goes on to look at the careers of four eminent scientists and to hear their views on ways to manage your career and your science. The last chapter (Chapter 22) provides a brief summary of the landscape upon which careers are mapped – specifically the higher educational system in the UK.

20

Managing stress

At the end of the day, your job doesn't love you back and everyone needs to have a good work–life balance to provide perspective and maintain sanity. To achieve such a balance, mental, physical emotional and spiritual aspects of life all have to be recognised, evaluated and given time. An A–Z directory of 'well-being' advice is provided here for consideration in the context of your own lifestyle.

The theory

Everybody knows that, in order to live a healthy long life, a balanced diet and regular exercise (both physical and mental) are essential. This recognition goes back to antiquity; the maxim *Mens sana in corpore sano* (a sound mind in a sound body) has been a favourite of many different people and organisations. Whilst there is a wealth of medical research published in medical journals, and some results have made it to the popular press, it would be irresponsible of us to recommend any particular exercise regime or any particular diet above others. The reason is simple – both have to be considered in the context of everything else in an individual's life. However, what is easily condoned is the idea that, to be in peak condition, you have to take time out of work on a regular basis – to look after your body and the mental, spiritual and emotional aspects of your mind. In effect, this is the best way to ensure that pressure (which can be motivating) does not become excessive and develop into stress.

The recreation and renewal of mind in order to be effective is acknowledged by Covey as the seventh habit of highly effective people (see Chapter 1). However, whereas Covey considers four distinct ways to refresh your mind – physical, mental, social/emotional and spiritual – we see them as interlinked. Moreover, there is no reason for classification – for some people baking a cake to share with others can be a spiritual activity whereas for others it is simply a gastronomic pleasure.

Well-being: a fusion between East and West

In this chapter we have gathered information about various approaches to 'well-being'. In gathering the information, it became clear that there is a difference between Western and Eastern approaches. In general, the Eastern approach is concerned with the circulation of energy through the body, and there is no such thing as a soul – body and mind are one. In contrast, Western approaches see a human as a body with a soul (for believers) or as just a body with a somewhat detached mind. As an extension, most Eastern techniques cultivate the mind through physical workouts, whereas most sports activities in the West aim at physical fitness only. Of course, this picture is a great trivialisation but it helps to understand some of the issues.

Currently non-religious courses offered in 'meditation and mindfulness' for well-being are gaining popularity. Intuitively, we know the meaning of these terms, but there are several definitions around. We came up with this: 'Meditation is creating breathing space, mindfulness is using this space consciously.' Another definition of meditation is *not-doing* as opposed to being active.

Not-doing is an anathema to a busy Western person. We are supposed to manage our time in the most efficient way, with some time management trainers advising people to schedule every 5 minutes. Keeping track of time is thus the ultimate achievement – if we have 'made it', we have no time.

The practice of meditation goes back to Shakyamuni Buddha, a well-known historical figure. Meditation has been practised for 2500 years in the East, and in the last century it became widespread in the West in various guises, of which Zen is one. The well-known example of how meditation can improve your health is that of Hakuin Zenji, one the most famous Japanese Zen masters of the seventeenth century (Yampolsky, 1971). Paradoxically, he first weakened himself with excessive meditation, became detached from daily life and very sick. Finally, he was advised by a hermit to practise in a different way and became very healthy and vigorous. He then spread the gospel in his teachings. We can safely assume that many monks and lay people alike practised that way over centuries and so this practice continues today. The question arises how this knowledge was made available to people in the West, who are generally suspicious of spending time apparently doing nothing?

In the second half of the twentieth century there was a breakthrough in medical science in America. Dr Herbert Benson, a pioneer of modern mind–body medicine, started to investigate the use of meditation in medicine. He studied the physiological effects of stress and demonstrated that the *relaxation response* can relieve stress more effectively that the *fight-or-flight* response. Factors such as metabolic rate, heart rate and blood pressure are increased by stress and decreased by meditation.

The importance of his research was to show a direct connection between the relaxation response and meditation. In his own words in an interview about the relaxation response (Redwood, 2008):

It is elicited by using two steps. The first is a repetition, which could be a word, a sound, a prayer, a phrase or even a repetitive movement. The second step is, when other thoughts come to mind, you disregard them and come back to the repetition. This would bring forth the same physiological changes that were brought about by the practice of transcendental meditation. The importance of this was that, again, for millennia people have been bringing forth a response opposite to the stress response, that has therapeutic value in disorders caused or exacerbated by stress.

Benson's work stimulated another seminal piece of work. In the late 1970s, early 1980s, a group led by Kabat-Zinn developed a programme at the Stress Reduction Clinic at the University of Massachusetts. Extensive research and work over 10 years with thousands of patients (and several physicians) demon-strated unequivocally the benefits of mindfulness and meditation in the treatment of chronic pain, anxiety and illnesses. Meditation training over 10 weeks with 90 patients suffering from chronic pain was highly effective in reducing pain and pain-related effects (Kabat-Zinn *et al.*, 1985). A spin-off of this work was a best-selling book with a programme for your own stress management (Kabat-Zinn, 2001).

The practice

Work–life balance

How does our personal ambition of being in top physical and mental condition square with our employers' aims? One of the main considerations of many employ-ers is the effectiveness of their workforce and the means to achieve it. Modern work ethics, particularly in the UK, are characterised by long working hours and by putting work before personal issues. This can easily lead to burn-out and is known to be counter-productive in the long term. Long working hours without adequate rest can significantly contribute to stress, especially if supplemented by fast food rather than by proper meals. Most institutions have advice on how to recognise and deal with work-related stress, and staff benefits such as access to sports facilities can help to maintain good physical and mental health. However, we each have to find our own balance (see examples in Box 20.1).

Pause for thought: Write your own A–Z of all the activities that you already do or think would be beneficial to you for relaxation.

Box 20.1
An A–Z of ways to relax
Alexander technique

This technique was developed in the 1890s by F. M. Alexander, a young Australian actor, in order to combat his problems with voice projection during performances. Nobody was able to help him and he nearly gave up his career until he noticed that his posture influenced both physical (voice projection) and psychological (stage fright) aspects of his performance. Correction of bad posture (e.g. hunched shoulders, cramped spine) through a series of exercises led to regained health and confidence. Randomised controlled trials of Alexander technique lessons have shown their effectiveness in alleviating chronic and recurrent back pain (Little *et al.*, 2008). The Alexander technique is now extensively used by actors and musicians, and is also helpful for public speakers (Gelb, 2004).

Bathing

Water has a very soothing effect as it helps to relax sore muscles – the beneficial effects of hot springs have been documented since Roman times.

Cooking

A rather underrated activity by many early career academics, but it can work wonders in three domains – physical, mental and social.

Drink

Moderation being the key!

Exercise

The benefits of exercise are so well known that we do not give any references.

Friends and family

They can be a great source of support (Riba *et al.*, 2007).

Gardening

Pulling out weeds clears the mind wonderfully for the next idea and helps create order out of chaos.

Holidays

Essential to have yearly to maintain your health and sanity – best without the laptop, mobile phone and email access.

Idleness

Sometimes, it feels good to just do nothing.

Juggling

As a form of exercise, it improves coordination and even increases brain-power (Jones, 2004). It is also a lot of fun.

Knowing

People outside of work – a good way to ensure that you talk about non-work related issues.

Laughter

A great and absolutely free device for getting rid of tension and getting back perspective.

Meditation

An ancient Eastern method for emptying (temporarily) the mind and purifying the spirit. While it is often associated with Buddhism, it is not in itself a religious activity. All meditation techniques are based on correct breathing and correct posture – the best known in the West are transcendental (TM) and Tibetan meditation. If regularly practised, it improves both physical and mental health and brings about increased energy.

Nutrition

Few subjects are as controversial, especially when connected with diet. In the UK, the official guidelines are by *NICE* – the National Institute for Health and Excellence.

Open mind

Helps to cope with new situations thus reducing stress in unfamiliar surroundings.

Pets

Keeping and caring for animals is known to be beneficial – most pet owners do not require any medical research to subscribe to it.

Qigong

A Chinese technique to develop qi (sometimes referred to as life energy). According to the adherents of qigong, strong qi and its free circulation in the body brings about health and longevity and keeps one balanced and centred. Often practised as a part of Tai Chi, an ancient martial art rooted in Taoist philosophy. Qigong is practised worldwide by millions of people.

Reading

Broadens your horizons – especially literature outside your own specialism.

Sleep

Its importance cannot be underestimated. Lack of sleep impairs effectiveness and can be the one of the first symptoms of stress.

Travelling

A change of scenery is often beneficial.

Unwinding

Consciously relaxing.

Volunteering

Helping others, contributing to the local community.

Writing

For fun, as opposed to a deadline for research article.

X-factor

A little extra something that makes you feel good – be it a piece of cake or an unplanned trip to the cinema.

Yoga

A form of physical, mental and spiritual exercise, with classes widely available. The benefits are so well known that we do not give any references here.

Zen

The teachings of this branch of Buddhist philosophy became much talked about in the Western world in the second half of the last century. Zen became an adjective symbolising simplicity (as in Zen minimalist approach), mind-baffling intellectual puzzles and mindfulness in everyday life. The core of Zen practice is meditation, often referred to as sitting (zazen), with emphasis on correct posture, breathing and concentration of mind. Its elements can be traced in every great religion and set of teachings (e.g. Confucius) as sincere prayer and reflection require concentration of mind. Zen is reported to be one of the most effective ways of concentrating and controlling your mind.

A final note on 'role stress'

Many academics suffer from role stress, the classic example being torn between research and teaching, and being resentful of administration. It is obvious that, most of the time, we play several roles at the same time, but often we only see one or two of the roles we play. Identifying all or most of your roles allows you to manage potential conflicts of interest, spell out expectations in each role, and to organise a weekly schedule to alleviate time pressure (Chapter 1). Awareness of these roles and being mindful of when we need to play them fully (i.e. when we need to focus exclusively on one role) helps to switch instantly from one role to another. Curiously, such mindful identification of our roles helps us to *dis-identify* from them as it allows us to realise that we are more than the sum of these roles. By not living out too many roles simultaneously, we master the art of living fully in the moment and this brings relief from time pressure – which, for most of us, is linked to thoughts about future actions.

How we did it

We tried two types of workshop. In the first, participants were asked to prepare their own A–Z dictionary of relaxation and we then had an in-depth discussion about the pros and cons of each method, and about the general need for making time to relax. In the second, one of us held a half-day workshop of talks on meditation and mindfulness, combined with practical yoga exercises. She has also piloted an 8-week meditation and mindfulness course for a small group of participants.

Summary

In this chapter we present information about how to combat stress in the form of a brief A–Z dictionary. We also describe research into the effects of meditation and mindfulness on well-being.

Selected reading

If you have time to read only one book, make it:
Kabat-Zinn, J (2001). *Full Catastrophe Living: How to Cope with Stress, Pain and Illness Using Mindfulness Meditation*. London: Piatkus.

Other texts:
Gelb, M. (2004). *Body Learning: An Introduction to the Alexander Technique*. London: Aurum.
Jones, R. (2004). Juggling boosts the brain. *Nature Reviews Neurosci.*, 5, 170.
Kabat-Zinn, J., Lipworth, L. & Burney, R. (1985). The clinical use of mindfulness meditation for the self-regulation of chronic pain. *Journal of Behavioral Medicine*, 8, 163–90.
Little, P., Lewith, G., Webley, F. *et al.* (2008) Randomised controlled trial of Alexander technique lessons, exercise, and massage (ATEAM) for chronic and recurrent back pain. *British Medical Journal* 337, a884.
Redwood, D. (2008). The relaxation response: interview with Herbert Benson, MD. *Health Insights Today*, 1, 1–5.
Riba, M. B., Riba, A. & Riba, E. (2007). Life as a balance beam: practical ideas for balancing work and home. *Academic Psychiatry*, 31 135–7.
Yampolsky, P. B. (1971). *The Zen Master Hakuin: Selected Writings*; translated by Philip B. Yampolsky. New York: Columbia University Press.

21

Taking on new challenges

Most academics become comfortable with their roles as researchers and teachers in their mid-40s. This still leaves 20 years before retirement, and there are several additional challenges that you can choose to take on at this point in your career. In contrast to earlier career stages, the choices are now mainly personality driven – you really can do what you want to do – as long as you do it well. In this chapter we present the personal stories of four eminent scientists to illustrate how varied your choices are.

The theory

It should be evident from the preceding chapters that a successful career in the sciences is based on several habits as defined by Covey (2004). It is not fashionable to talk about mission and vision, but all successful people have that – it drives their choices, and is supported by doing first things first. Further career enhancement often comes through working synergistically with others. If asked what contributed to their standing in the scientific community, many would say it was luck – they were in the right place at the right time. However, luck is a combination of preparedness and opportunity – if you are not prepared when an opportunity arises, or if you fail to spot an opportunity – you will be unlucky.

We could make a long list of the attributes that successful scientists have. But it seems that there is one characteristic common to many – having established themselves, they feel a bit restless in their mid-40s and look for other outlets for their energy. It is likely that their research is running smoothly, they are accomplished teachers and they can cope with their administrative load. So how can they direct their 'spare' time and energy into something different?

The practice

In what follows we present personal accounts as told by four eminent scientists. They all trained or work as physicists but their fields of expertise vary from astrophysics to materials science. Their career paths are very different, but they all pioneered new research approaches that later became standard. Two of these scientists are women. With more eminent women in science becoming role models for aspiring young female scientists, it is easy to forget that this was not always the case. Just a few decades ago, Professors Susan Jocelyn Bell Burnell and Julia Stretton Higgins (both Dames) were breaking the mould. Having achieved their recognition, they have since campaigned to increase the number and status of women in science, engineering and technology.

Jocelyn Bell Burnell DBE, FRS, FRAS is an astrophysicist famous for the discovery of the first four pulsars, while still a postgraduate student under the supervision of Dr A. Hewish. For this discovery he was awarded the 1974 Nobel Prize in Physics, jointly with Sir Martin Ryle *'for their pioneering research in radio astrophysics: Ryle for his observations and inventions, in particular of the aperture synthesis technique, and Hewish for his decisive role in the discovery of pulsars'* (http://nobelprize.org/nobel_prizes/physics/laureates/1974/). Jocelyn was not included as a co-recipient – a fact that many famous scientists have found disconcerting. She was President of the Institute of Physics (IoP) from October 2008 to October 2010, and at the time of writing is an Acting President of the IoP after the untimely death of her successor, Professor Marshall Stoneham. She is also a Visiting Professor of Astrophysics at the University of Oxford. In addition to being elected a Fellow of the Royal Society in 2003 and being made a Dame Commander of the Order of the British Empire (2007), she is the recipient of many honours granted by international organisations; for example, she has an Honorary Doctorate from Harvard University (2007).

Jocelyn's account of surviving and succeeding as a scientist has been edited from a transcript of a talk that she gave during a physics conference in Manchester in November 2010. (See also four video clips on http://www.open2.net/science/mscstudents/bell/bell_ind.htm#, last accessed 10/3/2011.)

> I started my academic career by failing the Northern Irish equivalent of the 11-plus, which must have given my parents pink kittens. I wasn't terribly happy either. In fact, I think I was probably more upset than I realised, because I didn't talk about it until I became a Professor of Physics, and thought, 'this is a bit silly, maybe I need to say something". I think part of the problem was bad teaching – because as soon as we started science in the next term at school, I came top of the class.
>
> In the first week of secondary school the girls were sent to the domestic science room and the boys to the science lab. The presumption was that girls would do domestic science, all of them, and all the boys would do science. My parents

actually protested loudly, as did the parents of a couple of other girls. And, in the next week there were three girls and all the boys in the science class. And, I would like to think that my coming top in the science exam at the end of that term gave the school pause for thought. But I'm not sure that it did.

This was the middle to late 1950s that I'm talking about. I come from a generation where women were expected to be wives and mothers, and to be entirely happy polishing the kitchen floor and doing that kind of thing. And that influences a lot of my particular pattern of work. It won't influence yours in the same way, but you may meet similar obstacles. So, I'm going to be talking a bit about problems and obstacles.

I did higher maths and physics at A-level and went on and did a physics degree at the University of Glasgow. They called it natural philosophy in those days, but it was a physics degree. Then as now, the Scottish physics degree was a year longer than the English physics degree. And, we all worked jolly hard. I was the only female doing honours physics. There were 49 men and me. Which meant you couldn't skip lectures, because it was pretty obvious. I once came top in a subsidiary maths exam and I thought they were going to lynch me. Women were for whistling at in that culture. And, I had a certain amount of that during my time in Glasgow. I graduated in 1965 with a BSc in Physics.

In my teens I had become interested in astronomy and decided that, if I could, I would do a PhD in radio astronomy. I was aware that, if I wanted to do radio astronomy as a career, I'd have to be doing it in a university. I'd be an academic, and that seemed quite a big step for somebody who's failed the 11-plus. So, all the way through I was keeping my options open. I did a physics degree, not an astronomy degree, because I thought, if I'm not good enough to be a professional astronomer, I need a job. I'll get a better job with a physics degree than with an astronomy degree. So, keep your options open wherever you can – it's quite important.

It's almost by chance that I got to do a PhD in Cambridge. I actually applied to Manchester, to Jodrell Bank. I was told my application fell down the back of the desk of a slightly chaotic academic who was doing the graduate admissions. But, I'd done a summer internship there and I'd been told by the grad students that they wouldn't take women. They had once taken a woman and she and a male student had put the dormitory to a use for which it wasn't intended. He bragged. Sir Bernard Lovell got to hear of it and said, 'no more women'. Interesting, which sex bears the brunt of these mutual encounters, isn't it?

So, when they didn't reply to my application, I thought this is 'no more women'. I didn't think I'd get into Cambridge, so I was gearing up to go to Australia. But the academic year in Australia begins January–February, so I had a few months in hand. Hence I put in an application to Cambridge just in case, and very much to my surprise got in. I really was immensely surprised.

I found myself in Cambridge, doing a PhD in radio astronomy and two big accidents of my life happened there. I was building a radio telescope for the first 2 years of my PhD, literally working in the field or occasionally in a very cold, draughty hut. The telescope was huge – it comprised 2048 antennae, more than 1000 wooden posts and 120 miles of wires and cables – and it covered an area of two and a half football pitches. So there were about six of us to build it. And, when it was built, the other five melted away and left me as a grad student to run this radio telescope.

There weren't many computers around. This was, by now, 1967. The head of the group had access to a computer as he was doing Fourier transforms. The rest of the academics had grad students instead of computers, and we analysed paper charts. And I had miles of it, literally, several miles of it, during the 6 months that I operated the telescope. Partly because I was still somewhat in awe of Cambridge, I was being very, very careful, and I came across something a little unusual, a peculiar signal.

Pursuing this unusual signal for once didn't lead to a dead end or something stupid – it led to a new kind of star, which today is called a pulsar. A pulsar is a very compact object, mass comparable to the mass of the sun, but a radius of just 10 kilometres. So it's very, very dense; comparable with the density of the nucleus of the atom. And this *thing* is spinning. And, as it spins, a radio beam comes out from the magnetic polar region where the field lines form some kind of funnel. And, as the star spins, the beam sweeps round the sky. Each time the beam swings across the Earth, you get a pulse. Pulse, pulse [you can hear the pulsar on the internet].

Pulsars – they're amazing things. They're highly dense. They are believed to be neutron stars. They were totally unexpected. Do you know when you write grant applications you have to say what you're going to research and what you're going to find? This idea would never be in anybody's grant application because we had no clue that anything as bizarre as pulsars existed. They were then considered totally ridiculous. And, they're still totally ridiculous today and it's very hard to believe they do exist. But I found the first four – we now know there are about 2000 of them. With a fingerprint of mass 10^{27} tons, 10 kilometres radius, spinning up to 700 Hz and magnetic field about 10^8 Tesla. If you spin it, then this gives you voltage drops of up to 10^9 volt/cm.

So, that was a bit of a diversion during my PhD, and I also got engaged to be married between discovering the second and the third pulsar. I nearly didn't make it with my fiancé to tell my parents we were getting engaged, because I found the second pulsar just the night before we were due to travel to Ireland. But occasionally you just have to stay up all night.

Both discovering pulsars and getting married had a huge impact in opposite senses on my subsequent career. I want to comment on that and what I learnt through going through that experience.

I married another Cambridge graduate who got a job in local government. In local government you get promoted reasonably fast by moving from one local authority to another every 7 years or so. We started in Hampshire, moved to Sussex, moved to Scotland – all over the place. And, after a few years there was a child as well, and I found society trying to put me in the role of a wife and mother. And, it was not a done thing for mothers to work. You'd hear male professors on the radio telling you that, if mothers worked, the kids were delinquent. There was consequently no child-minding facilities and no other support for mothers who wanted to work. It was pretty dire. There wasn't any maternity leave in the university when I became pregnant. Looking back, some things have improved in 30 or 40 years.

But, it was in this kind of ambience that I was trying to keep a career going, and it was jolly difficult. If the kid became sick, which kids do often, it was my problem because my husband had a meeting. And, because my husband kept moving, every so often, I'd write a letter saying, 'I'm moving to your area, dear director of observatory, or head of astronomy department, might you have a job for me? Preferably part-time.' And, I got the kinds of jobs that you get when you write begging letters. And, in the time that I was in any particular place I would gradually work my way up the hierarchy, and then my husband would say, 'it's time I moved jobs', and I'd go back down to the bottom of the hierarchy, find another job of some sort and work my way up again. It was extremely frustrating and very, very difficult. But it did have some plus points.

I did a PhD in radio astronomy. Subsequently I moved to γ-ray astronomy, to X-ray astronomy, to infra-red astronomy and to millimetre wave astronomy. I then got a professorship and was allowed to have my own research group, and we did multi-wavelength astronomy on neutron stars as the neutron stars remained my first love. Each time I moved, I judged it political to get involved in the kind of research that they did in each of these places. Of course, when you change fields a lot, there's a bit of an overhead in learning the new field and your publication record takes a dip. But, you end up with a very broad knowledge, with colleagues all over the place. You also become far too useful on research council committees because you can speak to proposals in virtually any wavelength band. But, having a broad knowledge of a field is very good, because I can go to a conference anywhere on anything and know the context.

My first postdoc was in Southampton University because my husband had a job nearby. I initially started doing work on the topside ionosphere, but quickly decided that wasn't quite right for me, and moved into the gamma ray astronomy group that was also in the same university. That money ran out. Second postdoc was also in Southampton – they had a lecturer leaving, they took the lectureship and they divided into two unequal halves and offered me the smaller half. And, I found myself with the normal teaching load of a lecturer, half the pay and none of the perks. But I did get to start lecturing as Junior Teaching Fellow in a university where they really cared about teaching – it was a great experience. I still occasionally have people contact me, saying: you taught me quantum mechanics, and I saw your picture at the IoP, or whatever, which is lovely.

Then my husband changed jobs again, and by this time there was a kid, so I went part-time to the Mullard Space Science Laboratory of University College, London. The best post I could get was as a part-time programmer, which meant I was a technician not an academic. I couldn't use the library without a letter from the professor . . . However, I had a great time there because X-ray astronomy was booming. It was really good. And then my husband changed jobs again and I ended in a job in the Royal Observatory, Edinburgh, which is outside academia, but within the Research Council ambit. I started briefly as a research fellow and then moved into a management position. So, by now, I had research and teaching and management experience. And, they were all good, brilliant; all useful. In addition, I have been a part-time tutor, consultant, examiner and lecturer for the Open University from 1973 to 1987.

Marriage broke up, kid went off to college, and I suddenly found myself for the first time in my life able to go after a job because of what it was, not where it was. I got the Chair of Physics at the Open University in 1991, and incidentally doubled the number of female professors of physics in the UK.

Let's pause here for a while – my last post in academia was as a technician and then I disappeared for about 10 years to re-appear as a professor. That's not a normal career path but, because the Open University recruited people on the basis of the skills they had acquired, what experience they had, I ticked very many of the boxes and I got the job. And, for the first time in my life I was my own boss, which was nice as well. Being Chair of the Department was quite a big management job so there wasn't a lot of time for research, but for the first time I also had my own graduate students. I stayed at the OU for about 10 years and then I moved to be Dean of Science at the University of Bath, which was a very big management job. Bath is very strong in the sciences and I was running about a third of the university. And then I 'retired', which I actually think is better described as another career phase where you do all sorts of crazy things like going around the country lecturing and being President of the Institute of Physics and saying no to things you don't want to do.

How did I survive without a 'proper' job? If you don't have a position at a university, there are a number of things you can do to keep going. Keeping in touch with colleagues, former colleagues and fellow students is very important. Equally important is to have a professional address. Don't use a private address as your only address. You lack credibility and you might be a crank. Ask the university you are having to leave if you can continue as an Honorary unpaid, visiting whatever and can you use their address and their email address. That's hugely important. Squatter's rights, if you like. You can do refereeing for journals, maybe even be an editor, help somebody with organising a conference, etc.; just anything to keep your toe in the door. Try to serve on a professional body and research council committees. Look for tutoring/part-time lecturing, be a lab demonstrator; help to supervise final-year student projects, offer your services with project reports, etc. It may be just pin money but you can hang in there. The big up-side is that you are getting many skills that can go on your CV.

However, don't just focus on your science and your academic stuff. You can pick up skills in other ways and you can articulate them. I reckon that, by running a home and raising a family, I got experience of budgeting. If you include the mortgage in the sum, you're handling quite a big budget. With kids, you learn time management and how to multitask. I've got to finish writing that paper by tomorrow morning, the kid's going to need my attention, but maybe I can get the kid down for a sleep and do an hour and a half and then maybe I'll do the rest at night. Things like that. And you learn a lot of interpersonal skills – negotiation, delegation. Persuading a toddler to do something is much, much harder than persuading your head of department to do something. So, identify these skills and put them on your CV.

My advice to early career scientists is to take risks because you may well surprise yourself. I think women in particular tend to be a bit cautious, maybe too cautious – just do it. Do not agonise over a failure – one failure does not make a disaster, be it at 11-plus or whatever – so treat failure as an opportunity for learning. Be persistent once you set your goal and aim as high as you can. Be open to new ideas and opportunities. Trust your instincts, and understand what your personal strengths are. I've learnt my instincts are a good guide, I have also learnt something they don't teach you in academia. Somebody asks a question in academia and you're trained to answer it. Stop and ask why that question is being asked? What is that person trying to achieve?

What's going on here? Watch the processes. And, finally, consider taking advantage of development courses offered by your university to broaden your experience.

There is a lot of exciting research going on in astronomy and astrophysics, and I expect it will go on for a while. Large-scale structure and the history of the Universe will figure prominently. On the instrumental side there will be big satellites launched (Gaia) which will map the Galaxy, and I expect there will be a successor to the Hubble telescope as well as a new generation of infra-red telescopes.

Jocelyn's story demonstrates intelligence, persistence, pragmatism and positive thinking, all traits that to a greater or lesser extent feature in the careers of most successful scientists. We asked three other eminent scientists to look back on their careers in the context of the following questions:

1. What do you consider to be your greatest achievement in science?
2. What was the most important factor in getting you where you are?
3. Why did you choose a career in science?
4. What do you feel have been the highest and lowest points of your career to date? How did you deal with them? How have they influenced your subsequent actions?
5. Did you feel you had to sacrifice anything in order to get where you are?
6. What was guiding you through your career choices?
7. Do you think you have/had a work-life balance, and how did you manage it?
8. In what ways do you think science as a discipline and as a career has changed over the last 30 years?
9. With hindsight, what advice can you give people starting their career in science now?
10. Given that advice, would you personally have done anything different, given the chance to try again.
11. What do you think the major scientific goals of the latter half of the twenty-first century will be?

Julia Stretton Higgins DBE, FRS, FREng is a Senior Research Investigator in the Department of Chemical Engineering, Imperial College, London. She was offered this post after her retirement as Professor of Polymer Science (1989–2007) and Principal of the Faculty of Engineering (2006–7). Her research studies the behaviour of complex materials, particularly polymers, in terms of their molecular structure, organisation and motion. One of her specialities is the application of neutron scattering techniques to such studies. Her current focus is on mixtures of polymers and on interfaces between polymers – either observing interdiffusion to understand bonding and adhesion of composites, or looking at the activity of copolymer additives as interfacial agents to improve properties of blends of thermodynamically immiscible polymers. She was the first woman to become both a Fellow of the Royal Society and of The Royal Academy of Engineering. In the Royal Society, she held the role of Foreign Secretary and Vice-President from 2001 to 6. She has had many other high profile appointments such as being Chair of the Engineering and Physical Sciences Research Council (EPSRC) from 2003 to 2007. In 2006 she joined the board of the Lonza Group, a company making chemical intermediates and biopharmaceuticals.

Julia talks about her prolific career:

I think my greatest achievement is being able to do an experiment, combine it with theory and translate the results into useful information. Specifically, I was able to translate different neutron scattering techniques into really useful information for people who are interested in dynamics, blends and similar things. This amounts to really trying to understand what the experiment is telling you.

The most important factors in my career success have been my personal curiosity and the input of a number of influential people, not least my physics teacher at school. I had a really good physics teacher when I was 15, without whom I would not have done science. Up to that point I had only been taught chemistry and I hated it. I knew I would not be a successful mathematician – I think I knew even then that I am an experimentalist.

I chose a career in science by default. At some point I did talk to a couple of people about working for industry, but I can't say that anybody advised me to do anything – there was no careers advice when I was at university. I fell into doing research and I really enjoyed it, so I stayed.

It is easier to pick the highest, rather than the lowest, points in my career. When your lectures go well or you get elected to the Royal Society, it is a terrific feeling. Taking on the jobs of Dean of City and Guilds College and Director of the Graduate School of Engineering and Physical Sciences at Imperial College were also high points.

It is quite difficult to think about a single low point. For a number of years I had no research grants awarded by the research councils – they were scoring just below the funding cut-off. However, I had departmental PhD studentships and also a little money from industry – so I had research students and kept going. It was not a lot, but enough. It was quite hard to run a group of eight PhD students with a wide range of projects and with no postdocs, and I did consider moving to different universities – but that was not necessarily going to help the grants get funded, so I stayed put. So, it was a low point, but it was not a disaster – I just used a pragmatic approach and did not give in.

Some of the low-lights related to management jobs – I hated running the department. Part of what you have to do in managing a department is, to me, boring. You have to understand finances and you have to deal with every person in the place. And it is all so repetitive – the same things come back every year. You do promotions one year, then you do promotions the following year and then you do them again . . . It was the same when I ran the faculty for a year. I agreed to do it and I had the total support of all the department heads but I still did not enjoy it. As far as I could see, you spend a lot of time dealing with problems that are not of your making and over which you have limited control.

I suppose the lowest point of all was when I was teaching at a school – I was really not cut out for it at all. I had no idea what to do at the end of a PhD, so I went into teaching – but I really missed research. I knew the moment I got back into research as a postdoc in Manchester, that I was where I belonged – people were paying me a salary for doing what I loved doing!

On a slightly different theme, there were times when I was a Reader at Imperial, before I became a Professor, when people were saying to me 'would you be interested in going to another university'. I looked at it, but people forget that, in the late 1980s there were a lot of cutbacks in science funding. And I remember saying to myself 'if you're on a big rock when the water is rising, you would be darn stupid to go onto a little rock'. This was an argument to stay at Imperial – which is not quite the same as high and low points.

I deliberately didn't sacrifice anything in order to get to my position – I worked very hard, but it did not feel like a sacrifice. I remember saying to myself at some point 'What will I regret not doing in my life?' I realised that it was never going to be writing another paper – this gave me a sense of perspective and priorities.

What was guiding me during my career? I think that you have to be lucky, but you have a role to play in your own luck. A few times I took what some people thought was a mad decision – I did something unusual. For example, when I was finishing my physics degree, I always said that I wanted to do research in elementary particles – everybody wanted to do elementary particles at that time. I was offered a studentship at Imperial in elementary particles, but at the same time my tutor introduced me to John White who offered me PhD in Chemistry at Oxford University. I hated chemistry, I knew nothing about it, yet I chose to do that PhD. If you think about it – if I had not moved into neutron scattering – my whole career would be completely different. It was a very good decision, but it was surprising – it was not what I wanted to do, but somehow I recognised it as the right thing to do. Of course, part of this recognition process was that I wanted to stay in Oxford, with people I knew – but I was moving into completely different field. The subsequent postdoc in Manchester was just luck – I wrote a lot of letters and ended up there. Again, however, a lectureship in Chemical Engineering was not the obvious next thing to take up – but it was a good thing to do. I think, at the end of the day, it is crucial to be aware of what you want and to have self-knowledge. I think that women are better at knowing what they want than men, and they admit it. They undervalue themselves, but actually they understand themselves better than men. Men go leaping after what you may call 'glory', forgetting that there is something else that they really want to do.

It surprises me that postdocs consistently avoid participating in activities that encourage you to understand yourself. It does not need to be deep psychology, it can be fun things like working in groups – but you need self-knowledge to understand what you want to do. And then you need some courage if an opportunity looks peculiar – and you need some luck!

As for work–life balance, I tended to separate the two. I very rarely worked at weekends because I saw my partner then; the weekends were our time. So, in the week when he was not there, I was very organised – I would come to work at eight in the morning and leave when things were finished. I notice that, now we are both retired, things are much more muddled up together!

There were times when I got stressed – really stressed – normally when I was under time pressure and in a management role. You have meetings all day and then you get told to return comments on an EPSRC proposal in 4 days. Sometimes, it just can't be done and, after a while, I realised that you have to be ruthless and say that you just cannot do it. You also have to say no to more things in general.

Looking over the last 30 years, science as a discipline has changed. I think the pressure on young lecturers is much higher than it was when we started. It would be very difficult for somebody to survive the way I did. Not least because the system is running on the overheads you bring in from research grants. I got my first PhD student after a year or two because there was one available. I don't remember being under pressure to do that, I just wanted people to work with me. There is also less time to reflect.

There is also much more financial pressure if you're working in any of the big cities because of the house prices. Salaries are not bad, but it is still very hard to run a family home on a lecturer's salary, certainly in the near vicinity of a university, and so people are commuting long distances, which wastes time and energy. I am full of admiration for young women who are coping with a career and a family – it is difficult to see how they do it.

To recap, I would advise people starting their career in science that it is crucial to learn about yourself and your skills – and not just your technical skills. Don't sit glued to your desk all the time – be outgoing, make as many connections as possible, do courses on teaching, go to conferences, meet people and so build your networks – you never know when somebody may come back and help you. And then, I think that being flexible and open-minded will allow you to see opportunities that are not immediately obvious – don't be afraid to take a slightly peculiar opportunity when it comes along if it feels right – but if you don't want to do it, then don't.

Given that advice, I would personally not have been a school teacher. However, in recent years it has been quite useful to have this experience when talking to schools and educationalists. I should also have done some chemistry in the sixth form, but then I would not have done so much maths . . . It sounds terribly smug, but I'm very happy with my career.

In terms of the major scientific goals for the future, I think that these will inevitably be associated with the grand challenges to humanity – energy, water, food. That is, how do we keep ourselves comfortable in an increasingly crowded world with diminishing resources.

An eminent particle physicist, **Peter I P Kalmus OBE** is an Emeritus Professor at Queen Mary, University of London, and an Honorary Fellow of the Institute of Physics. Over his illustrious career, Peter has been the author of more than 230 research publications and has taught students at all levels. He has been particularly active at the Institute of Physics over the past years, serving as a member of Council from 1993 to 2000 and as Vice-President from 1996 to 2000, with special responsibility for education and public affairs.

Peter has always been interested in furthering the public awareness and understanding of science. He started giving outreach lectures to Women's Institutes in the late 1950s, initially on nuclear power (then a novel source of electricity). At that time, the popularisation of science was regarded with suspicion by some senior members of the scientific establishment. However, Peter always felt that scientists had a duty to explain what they were doing, and persisted throughout his university career. It therefore amused him when an influential report from the Royal Society came to the same viewpoint a quarter of a century later. When the new UK Research Councils came into existence in 1994, such outreach activities became not only respectable but almost mandatory.

Peter is well known for his popular lectures to schools and general audiences. These have also been delivered at meetings of the British Association for the Advancement of Science (now called the British Science Association), at the Royal Institution of Great Britain, the Institute of Physics, the Association for Science Education, the Edinburgh Science Festival, the Bournemouth Arts Festival, to science and astronomy clubs, and abroad in several continents. He has made a number of short radio broadcasts on scientific topics, and has appeared occasionally on TV. In recent years he has given 200 talks on particle physics to about 30 000 school pupils and others in the UK, Ireland, South Africa and India. Peter received the Institute of Physics Kelvin Medal and the European Physical Society Outreach Prize for this work.

Peter talks about his illustrious career:

> I think my part in the discovery of the W and Z particles in 1983 was my greatest achievement. This was the experimental verification of the electroweak theory of particle physics, which unified the electromagnetic and the weak forces. The discovery was made at CERN's proton–antiproton collider. We were very happy when the two people who made the biggest contribution to this adventure, Carlo Rubbia and Simon Van der Meer, received the Nobel Prize. More than 100 people were involved in this experiment, several received prizes from scientific societies, and I was fortunate to receive the Rutherford Medal and Prize for my part in this discovery.
>
> The most important factor in getting me where I am now (Emeritus Professor of Physics but only notionally retired) is probably just luck. The situation in the 1950s when I got my BSc was so fantastically different from now. I think the expectations of people in my age group were not very high after the Second World War, but the opportunities for employment were good. From about the mid 1950s there was a great expansion in UK science: physics, in particular, was held in high esteem as it was thought to have contributed to winning the war. I was offered a postgraduate studentship at University College, London and 3 years later was offered a postdoctoral job there without actually having to apply. They were looking for people who would be able to do research. Afterwards, I spent a few years in America and was then invited to take up a Lectureship at Queen Mary, again without having to apply for it. I had been recommended by someone and things went from there, so perhaps this is not useful information for today's scientists as they have to fill in many application forms before getting a job.
>
> Apart from these lucky openings, I was fortunate with my PhD supervisor whom I thought was excellent. I was in almost daily contact with him because there were four

of us who were building an accelerator, so he really wanted to know what I had done and whether it was the right thing to do. While he steered me in the right direction, he allowed me to make mistakes and that's quite useful, so I had a good start. As a postdoc, I again worked with somebody very good. Generally, I've been very fortunate with the people I have worked with – initially my bosses and afterwards my colleagues and my students.

My father was a scientist, a biologist, so the scientific process was just something I grew up with. He had a particular interest in the genetics of the senses – vision and hearing, etc. – so all family members (who are genetically linked) were automatically subjected to the Ishihara test for colour-blindness and relevant tests for the other senses! When I was 6 or 7, my father showed me around his lab and at that point I realised that there were such people as scientists. At school, I was good at maths and science, but in our particular school we had to choose specialisations rather early. As it was wartime, there was a shortage of teachers and we didn't really have a biology teacher. So I did physics and chemistry – and physics was the one that I enjoyed most.

The highest point of my career was the work at the proton–antiproton collider at CERN. That was a world-class set of experiments and it was very exciting to be at the forefront of science. We were able to use the collider, which had much greater energy than any previous machine. More or less anything that we did with that machine was new and exciting, and discovering the W and Z particles was the highlight of our research. We probably got about 60 publications over the 10-year period of that programme.

I don't think I had any low points in the sense of being desperate about my career, but there have been disappointments. There were a couple of experiments that didn't give very good results – that was a disappointment. Also, when I was Head of the Physics Department at Queen Mary College, I very much hoped that our department would get the top grade in the Research Assessment Exercise (RAE), but we didn't. I knew how to deal with it – but it took a long time to persuade the college management what was needed. In the next RAE we did get the top grade, but I was no longer head of department – that was a bit sad after all the effort I put in.

I don't really know what has guided me through my career. I worked hard but I enjoyed it – in a sense physics is my hobby as well as my career, so I always felt I was very fortunate to be able to have a job doing what I like and being paid for it. I didn't have very many hobbies, but I travelled a lot, so I didn't feel I had to make the sort of sacrifice that many female colleagues have to make. They have to put in the same amount of effort as their male colleagues at work, but normally have more than half the share of home duties. My wife has been very supportive – while our life was somewhat restricted by her state of health, I don't really feel I had to make any significant sacrifices. Probably, my work–life balance wasn't quite right, but not really for physics reasons.

Another thing that influenced strongly my work–life balance was working in collaborative teams at large facilities. These accelerators used to run around the clock, so we did shift work like in wartime factories. It was quite a common thing to have 8-hour shifts and to work every fourth one, so that meant quick adjustments to a different time. And, of course, mixing that in with teaching, which was always in a different location, wasn't so easy. All of the work that I've done during my career has been collaborative, starting with rather small groups and ending up with very large groups. You had to have some sort of consensus of what to do, and what the priorities were, and sometimes there would be differences of opinion. I think that most of the decisions we took were reasonably sensible and it may well be that sometimes when I disagreed, then with hindsight, I might have been wrong. I enjoyed working with other people – it was particularly interesting to do that in an international setting with people from several other countries who had different national characteristics. I mean we're all scientists, but there are some national differences in how people do things and in their background. Some of our colleagues from other countries were a lot better at maths than I was, because they'd had a much more rigorous training, and they might not have been quite as good at some of the experimental techniques as I was – but this meant that within the collaboration we had many areas of expertise.

If I look back to where I started from, there have been immense changes in science as a discipline and as a career. Turning first to the latter, I was very fortunate to be in an expansionist period in the early part of my career. Things are improving now, but people who were 10–15 years younger than me had a much harder time. There was a great expansion in the 1960s and early 1970s in university physics departments – and then they were full of young people. For many years, until they retired, it was just the same people getting older and older, so it was very difficult for young people to obtain permanent university jobs. Fortunately, there were many other jobs outside universities for our graduates, postgraduates and postdocs, particularly in computing, and they tended to pay better than universities.

I think, in terms of science as a discipline, there have been several moves. Firstly, there's been a move towards inter-disciplinarity – there's always been some, of course, but over the last several years it has become much more pronounced. This has led to much more group work – I'm not just talking about particle physics where the author list is sometimes longer than the paper – in many of the physical and life sciences it is now quite common to see multiple author papers. Secondly, science has become more bureaucratic – there is a lot more writing applications for grants; general bureaucracy both within a university when applying for resources and on the outside, dealing with research councils or international organisations. Probably, in my area of experimental particle physics, it is not so bad because new lecturers don't initially have to get their own finance because we have a rolling grant, we have a programme. Others have it tougher – I've been on selection panels for people in other sciences and the first thing they were asked was "where are you going to get your support from?' The ones who get the jobs already have contacts with outside companies or can transfer a grant form their previous institution. Finally, because of the general expansion of science, people at the top are much more remote than they used to be. When I first came to Queen Mary as a Lecturer, I would occasionally sit next to the Principal of our university at lunch and have an informal chat. Similarly, I could have coffee or lunch with the Director of the national laboratory that we used. I've only spoken to our present Principal on one occasion – I'm retired so I don't have any real need, but I'm in every day. He is 'up' in an office and there are several layers of people between him and me, which seems fairly common elsewhere too.

I think science is brilliant. It generally requires hard work, but my advice to people starting their career is to go for it. There are many interesting things that can be done. It would be very good if more people studied science, firstly at school and then at university, even if they didn't become scientists. It would also be good if some people who have carried out research for a certain number of years would then move onto something else. We need some scientifically trained people in all walks of life. We're absolutely surrounded by the fruits of science, and the technology based on science. Citizens have to make informed choices on all sorts of controversial technical things and that's very difficult for people who don't have a scientific background. For example, it would be very good if we had a number of Members of Parliament who'd been scientists for a while; we've got one or two but that's all, very, very few. At least in the House of Lords there are a number of people who know about science.

Would I personally have done anything different, given the chance to try again? I don't quite know how to answer that because my advice is not very specific. I'm just saying 'go for it' because science is great. It is interesting and also useful to society. I don't think anybody actually gave me that advice.

Given my record, I'm sometimes asked about public outreach, when to start it and how to balance it with the rest of your duties. I started when I was a postdoc and the Head of Department got a request to find somebody to give talks on nuclear energy to Women's Institutes. I think what happened, though I can't be sure about it, was the request went down the hierarchy and there was nobody lower than me, so I was volunteered. And I didn't quite know what to do – in those days (in the late 1950s) we didn't have anything like Powerpoint and data projectors (we didn't have computers) or even optical overhead projectors. I had to talk to intelligent but most likely scientifically illiterate people. At that stage, you had to tell them, as far as the consumer or the user was concerned that nuclear power just came out of the wall socket, it was electricity. I didn't give a huge number of outreach lectures, but none the less I did them all the time, a few per year maybe. I somehow felt that, having been funded by the taxpayer, it was my duty to tell people about science if they wanted to listen. After I retired, that became one of my main activities – research was what I enjoyed most during my career, but when I got to retirement age I was in a collaboration of 400 people working at a collider in Hamburg. I decided that in retirement I didn't want 399 collaborators. Unless I worked flat out, I would probably feel like a passenger. I can do outreach work on my own. I don't even need much in the way of resources, although, because it has become popular, you can now get grants to do it and get prizes for doing it. It has become not only fashionable but almost compulsory.

I am reluctant to speculate on the major scientific goals of the latter half of the twenty-first century, as I'm fully aware that predictions of this type by people who are a lot cleverer than me have nearly always been wrong, and sometimes dramatically wrong. For example in the 1940s the head of IBM estimated that the world would need only five or six computers and in 1954 the Chairman of the United States Atomic Energy Commission predicted that electricity would become too cheap to be metered. I would have thought, in the nearer future, it's likely that there would be more emphasis on direct applications of science to the needs of society. It seems probable that climate change, sustainable energy, population growth, use of finite natural resources and various problems connected with health and ageing will require additional research, but I hope that fundamental science will not be neglected, because that is where the breakthroughs are likely to be made.

Adrian Sutton FRS is Professor of Nanotechnology and Head of Condensed Matter Theory in the Department of Physics, Imperial College, London. His research interests are at the interface between condensed matter physics and materials science – he applies fundamental physics to understand and predict the structure and properties of materials of technological significance. Adrian has frequently developed new concepts and undertaken pioneering simulations of materials at the atomic and microstructural levels that have made a lasting impact in materials science. His research is theoretical and computational in nature, and is of a problem-driven rather than technique-driven nature, spanning as it does metals and alloys, ceramics, semiconductors, polymers and composites, functional and mechanical properties. He is one of the most highly cited materials scientists worldwide, and his contributions to materials science were recognised in 2003 through election to a Fellowship of the Royal Society. Since 2008 he has been Director of the EPSRC Doctoral Training Centre on *Theory and simulation of materials* at Imperial College. He is the recipient of several honours and awards, the most recent being a Royal Society Wolfson Merit Award (2006–2011).

Adrian writes about his career:

> My greatest achievement in science? If you are asking about the research with which I am most pleased, then I suppose it would have to be either the work I have done with many people on the structure and properties of grain boundaries, or the work I did with Tchavdar Todorov on relating the mechanical and electronic transport properties of metallic nanowires. The research monograph *Interfaces in crystalline materials* I wrote with Bob Balluffi over a period of 6 years is the publication with which I am most satisfied. This book is a synthesis of a vast area at the heart of materials science, and it set out the fundamental science of interfaces rigorously for the first time. It is certainly the largest work of scholarship I have been engaged in. When I open it today, 16 years after it was completed, it can still send a tingle down my spine.
>
> If I interpret your question as asking what it is in my career as an educator in science with which I am most pleased, then the answer is the EPSRC Centre for Doctoral Training (CDT) on Theory and Simulation of Materials (TSM) I have established at Imperial College. This is something I have wanted to do ever since my own PhD at the University of Pennsylvania 30 years ago. The Centre is my vision. I wrote the proposal, I became the Director and attracted superb individuals within Imperial to help me run the Centre and deliver the courses I designed, and I have set the central research theme of the Centre. Imperial has been very supportive, but inevitably there have been political battles that needed to be fought and I have fought and won them. I have persuaded more than 60 academics across six departments in two faculties to get involved in the delivery of the teaching, research and outreach of the Centre. But the greatest success of this Centre is undoubtedly the quality of the students the Centre has attracted from the UK, the EU and overseas. Never before have I seen so many brilliant students eager to learn about theory and simulation of materials. I see many of these students eventually becoming world leaders of the field.

It is difficult to single out the most important factor in getting me where I am now. My father was an industrial chemist before he became a chemistry school teacher. His passion for science was a strong influence on me from a very young age. My physics and maths teachers at school were very encouraging and inspiring. Going to the University of Pennsylvania for my PhD had a decisive influence on me. At Penn I was able to spread my wings and take superb, advanced graduate courses in physics, maths, mechanical engineering and materials science. Those courses have had a lasting impact on my career, and they shaped my vision for the CDT I established 30 years later.

In 2004, when I decided to leave Oxford, it was my wife, Pat, who made me realise I could and should apply for the post at Imperial I now hold. Without her encouragement and support, I doubt I would have applied.

The only decision I can recall making about a career in science was when I had to select my A-levels. It was based on a hard look at my academic strengths and weaknesses. After that, I just took opportunities as they arose. If I rephrase the question to 'what is attractive about a career is science?' I would say (i) it is one of the most creative of all human activities, (ii) it is an intensely human activity, by which I mean you invariably work collaboratively, (iii) it is the only way to solve many of the largest problems humanity faces.

I believe I am at the highest point of my career now, directing the CDT, and heading up the Condensed Matter Theory group at Imperial. In 2006 I was one of four people in London who founded the Thomas Young Centre, the London Centre for Theory and Simulation of Materials, which has already become very well known internationally. I have taken a leadership role in building up materials physics at Imperial, and London more generally, to its current world-leading position in education, training and research in TSM.

The lowest point of my career occurred during the last 2 or 3 years of my 23 years at Oxford. I had come to feel undervalued, overworked, undersupported administratively and miserably paid. My disillusionment with Oxford began when I reluctantly agreed to apply for a titular professorship in 1997. The terms of the post were that there would be no change of duties and no change of salary, but only the latter turned out to be true. My morale reached an all-time low when the University pay deal for staff like me who were not tutorial fellows required me to double my tutorial teaching load to receive a pay increment of just £3000 per annum. I failed to make the Vice Chancellor see the injustice of this deal. My own sense of self-worth had been so diminished by Oxford I felt trapped because I didn't think anywhere else would want to recruit me. This is perhaps even more remarkable because it coincided with my election to a Fellowship of the Royal Society in 2003, before my 48th birthday. After I left Oxford, my post was converted from a lectureship to a professorship with a professorial salary, presumably at a cost of significantly more than £3000 per annum. When I took up my post at Imperial in January 2005, my pay literally doubled. I think the most important point that I learnt from this whole experience is that you can be too long at any institution. One should always move on rather than try to fight the institution. I should have left at least 5 years earlier than I did.

In order to get where I am, I can recall two personal sacrifices. When I went to Philadelphia to do my PhD, I left behind a girlfriend with whom I was very much in love. My first year at Penn was very difficult, but eventually I reached some sort of accommodation. When I returned after my PhD, we had both changed so much that we eventually broke up, which was again very painful. But, I subsequently met my wife and I have never looked back with any sense of loss or longing. I decided to go to Philadelphia despite the personal sacrifice because I really wanted to work with my supervisor, Vasek Vitek. I felt it was a great opportunity scientifically, and now I know I was right. The other personal sacrifice was taking a joint appointment with Helsinki University of Technology (HUT) in 2003–4. This was the only way I could find, at that time, to reduce the time I had to spend at Oxford, where I was so unhappy. But my family remained in Oxford and I commuted to Helsinki 12 times in the summer of 2003 and more than 20 times in the summer of 2004. It was exhausting, but I enjoyed my time scientifically and culturally at HUT very much indeed.

Achieving a work–life balance has never been one of my strengths. I work far too much, as my wife keeps telling me. But in the last few years, Pat and I have had to deal with ageing parents and securing the support and care they need. This has led me to rethink my priorities, and from October 2011 I shall become a part-time employee at Imperial to create more time for Pat and me to have some fun together.

The working lives of academic scientists have changed enormously over the past 30 years. Some of my heroes in science who were active until the 1980s never had a grant and never had to apply for one. Many senior academic scientists would describe themselves now as much as business managers as scientists. Keeping a large group funded has become almost a full-time occupation by itself. There is also far more top-down management in UK universities, as central administrations coordinate their responses to the RAE, teaching quality assessments, audits by funding agencies, national and international league tables, and the fear of health and safety, employment, human rights and other litigation. Thirty years ago a head of department was a powerful individual – a baron or (very rarely) baroness in the university hierarchy. In my opinion they are now little more than functionaries for the institution.

Science as a discipline has also changed. Big science is much bigger and much more expensive. Computers today are thousands of times faster and more powerful than 30 years ago. Science has become a truly international activity, with collaborations across the globe and instant communication through the internet. Scientists move much more freely from one country to another now that the Iron Curtain has been removed, and the market for scientists has become truly international. The UK Government knows that, if it treats science badly, many of the country's finest scientists will take themselves and their groups overseas where they will be welcomed.

Science has always had fads. But, the reality behind the hype of some of them has exposed the danger of slavishly following these trends at the expense of less fashionable but nevertheless strategically important areas for the national economy. An example is the explosion of nanoscience and nanotechnology through the 1990s, which displaced the funding of so much core materials science, such as metallurgy. Metallurgy is no less vital today than it has ever been, but thanks to the funding patterns of the past 20 years there are now very few metallurgists in the UK. For many industries in the UK, this is nothing less than a catastrophe.

In my view the greatest challenge scientists face today is the same as it has always been – attracting the brightest young minds. The magnitude of some of the global challenges we all face has drawn many more bright young people into science, and they are our greatest hope for the future. What advice can I give people starting their career in science now? Like choosing a life partner, you should look around before you settle down somewhere. The funding of science, the social standing and pay of scientists, and the quality of the education and training students receive, all differ widely between different countries in the EU and beyond. You may find that you would be better off starting your career outside the UK. You are in an international market, so my first advice is make the most of it. If you decide to start your career in the UK, don't be in a hurry to get a permanent post. Permanent posts come with permanent responsibilities of teaching and administration in addition to the research you love. The most productive and creative period of my research career was the 8 wonderful years I held a Royal Society University Research Fellowship at Oxford. During that period I spent more than 95% of my time on research, and I was able to build an international reputation in my field. Speaking of starting up again with the benefit of hindsight, I have often wondered whether it was a mistake that I returned to the UK after my PhD in the US. But I have no regrets. The Royal Society looked after me superbly with my URF, and I enjoyed being a lecturer at Oxford between 1991 and 1997.

I have no idea what the major scientific goals of the latter half of the twenty-first century will be. But that's part of the excitement of science – it is completely unpredictable. All the major breakthroughs in materials science during my professional life came completely out of the blue. This is why I don't believe Government Foresight exercises have any value – Faraday would have spent his career refining gas-lights.

Summary

In this chapter we have presented the rich careers of four distinguished scientists. All of them faced obstacles at different times in their career and all wanted to do more than just research. They took risks, actively searched for fulfilment of their scientific ambitions and opened new fields of inquiry. They have made lasting contributions to science but are also model 'citizens of the republic of science', giving their time generously to lead and educate others. While they were fortunate to be in the right place at the right time at crucial points in their careers, they made their own luck by hard work, ingenious use of their talents and working with others.

Selected reading

Covey, S. R. (2004). *The 7 Habits of Highly Effective People*. London: Simon & Schuster.

22

The higher education system

To excel in research, teaching and learning, an academic scientist needs be aware of the educational context within which they operate. This chapter provides a brief survey of the higher education system in the UK. Although the details are specific to the UK system, many of the principles are equally applicable in other countries.

The theory

The state of education in any country is intimately linked to that country's history and to the national perception of how useful an education is for making a living. In the UK this picture is complicated as the education system is devolved: England, Scotland, Wales and Northern Ireland have their own governance and funding systems. In all four sectors, however, over 80% of the funding for teaching and research comes from the public purse. In England, the Higher Education Funding Council for England (HEFCE) funds 130 Universities and 124 Further Education Colleges (with a 2010–11 budget of £6.5 billion provided by the Department for Business, Innovation and Skills). The Scottish Funding Council (SFC) supports 16 Universities and four Higher Education Institutions (collectively known as the university sector) plus 43 Colleges, with a budget of more than £1.7 billion per annum. In Wales, 11 Universities and a number of Further Education Colleges are funded by the Higher Education Founding Council for Wales (Cyngor Cyllido Addysg Uwch Cymru) to the tune of more than £440 million. Seven Irish universities, 14 Institutes of Technology, 9 Colleges of Education, and a few other Higher-education Institutions are funded by the Department for Employment and Learning with a current annual budget of around £500 million.

The amount of public funding available for higher education varies (sometimes drastically) depending on the economic climate and on government policy. Despite this volatility, universities are expected to maintain both their constituencies and the consistency of their research, teaching and learning. Some of the differential between the sum required to run an institution and the funding received from government funding agencies (either through direct grants or research grants) thus has to be met by grants from industry and charities; by endowments and by income from student fees. According to data collected by the Higher Education Statistics Agency there were over 180 000 academic and 205 000 non-academic staff employed in the UK HE sector in 2009–10, and over 2.4 million students were registered.

The practice

All universities aspire to excellence in teaching, learning and research. In 1994, UK universities amalgamated into distinct groupings. The so-called 'Russell Group' of 20 research-led universities collectively receives two-thirds of all research grants and contract funding awarded in the UK. The Russell Group is sometimes referred to as the British Ivy League. Its members are:

- University of Birmingham
- University of Bristol
- University of Cambridge
- Cardiff University
- University of Edinburgh
- University of Glasgow
- Imperial College London
- King's College London
- University of Leeds
- University of Liverpool
- London School of Economics & Political Science
- University of Manchester
- Newcastle University
- University of Nottingham
- University of Oxford
- Queen's University Belfast
- University of Sheffield
- University of Southampton
- University College London
- University of Warwick

It is not just the quality of research that makes the Russell Group outstanding – there is also tough competition for undergraduate courses, with an average of seven applications per place offered. These universities also have the lowest dropout rates and some of the highest progression rates from undergraduate to postgraduate studies. In addition to the Russell Group there is the '1994 Group' that is composed of 19 smaller research-led universities, the 'University Alliance Group' of 23 business-focused universities and a university think-tank called 'Million+' that comprises 27 post-1992 universities (previously polytechnical colleges).

But how is the national and international standing of a university or a college assessed? The Quality Assurance Agency for Higher Education (QAA) was established in 1997 to assess how well Higher Education Institutions (HEIs) meet their responsibilities. A key part of QAA's work is to develop guidelines and benchmarks for establishing and maintaining quality across the HEI sector. A more 'consumer-driven' method of assessment is manifest in league tables published annually by a number of national newspapers. The aim of these league tables is to inform potential students about what they can expect at a given institution. The ranking criteria include entry standards, student satisfaction, staff:student ratios, research quality, services and facilities spend, competitiveness, honours bestowed and graduate prospects. Of these criteria, research quality and student satisfaction (related to quality of teaching and learning) are directly correlated with the performance of the academic staff.

Research quality

For many years research quality has been assessed in UK universities through periodic (every 5–7 years) Research Assessment Exercises (RAE). In its latest incarnation, the Research Excellence Framework (REF) in 2014 will:

- inform the selective allocation of research funding to HEIs
- provide benchmarking information and establish reputational yardsticks
- provide accountability for public investment in research and demonstrate its benefits.

Over the years the details of each assessment have varied, but the aim has always been to produce a 'quality profile' of research conducted across all HEIs, so that funding for any particular HEI reflects its level of national and international standing. These assessments are taken very seriously by universities because the result influences the amount of public funding awarded to the institution, which, in turn, influences the ability of the HEI to attract high-quality staff and students and thus to attract other sources of funding. This pressure is felt by individual academics because everyone's contribution counts – both in terms of the amount of grant income secured and the quality of papers published. Needless to say, this pressure is not particularly welcomed, and is seen by many to distort academic pursuits. However, such assessments are undoubtedly here to stay and it is important to be aware of the criteria that are being used to measure 'quality'.

Teaching and learning quality

In most UK universities it is compulsory for new lecturers to follow a professional development programme when they start to teach. Completion of these courses merits a certificate or diploma, and many lead to recognition by the Higher Education Academy (HEA). HEA is a network of discipline-based subject centres with a mission to provide students with the highest quality learning experience in the world. Working with universities and individual academics, HEA operates a professional recognition scheme on three levels: Associate, Fellow and Senior Fellow. The first two grades can be awarded either through courses accredited by HEA at universities or through an individual route. Awards are recognised across the UK HE sector as a measure of teaching excellence. Each year HEA also grants up to 55 Individual Awards of £10 000, under the auspices of its National Teaching Fellowship Scheme. These awards recognise personal excellence and are intended to fund further professional development in teaching and learning.

'Home' students

To fully understand the HE education landscape in any country, you need to understand the educational background of the student cohort. For anyone who did not go through the school system in that country, this is not always a trivial matter. For example, in the UK there is often confusion between the types of school: public vs. state; selective vs. comprehensive; mixed vs. single sex; secular vs. religious. Paradoxically *public school* refers to a private (or independent) school that is funded through tuition charges, endowments and gifts (so is not funded by public money). *Selective* schools accept pupils only if they meet certain criteria, mostly academic, while *comprehensive* schools accept all pupils.

As with higher education, primary and secondary education in the UK are devolved in Scotland, Wales and Northern Ireland. There is compulsory schooling from the age of 5 to 16 (referred to as years 1–11) and then optional schooling from 16–18 (Years 12 and 13 or 'the sixth form'). At any one time ~75% of 16–18-year-olds are in full-time education at schools, Sixth Form Colleges and Further Education Colleges. Education is free for anyone, but many parents/teenagers opt for fee-paying schools.

Although some schools offer the European and International *Baccalaureate*s (EB and IB), the normal qualification milestones in UK secondary education are General Certificate of Secondary Education (GCSE) and Advanced (A) levels. GCSEs are studied over 2 years from the ages of 14–16 and students normally study between five and ten subjects. GCSE grades are awarded from A* to G/U is unclassified, and in order to progress to A-levels there is normally a requirement for 'good GCSE passes'. A-levels are also studied over 2 years from the ages of 16–18, with students normally taking between one and four subjects. The grades awarded again range from A* to E, and different universities have different specifications for entry, both in terms of the number of subjects and grades required.

Prospective students apply to study at a university or a college through the University and Colleges Admissions Service (UCAS), by filling in an online application form with their choices of institution(s) and course(s). These forms are then sent to the relevant university departments, some of which interview potential students. In terms of the academic strength of applications, the 'UCAS Tariff' allows university admission tutors to compare a range of different qualifications (EB, IB, A-levels). However, other factors are also taken into account and admissions requirements vary widely between different universities and between different courses. Notably, the Open University accepts all students regardless of prior qualifications.

'Overseas' students

In most HE institutions, the student cohort also comprises overseas students. They come from a different culture, may not speak English well and have different expectations of a university education and of a teacher's role. In 2009/10 16% of students in UK HE institutions were from overseas and there are indications that this proportion will increase in the future. Other countries are registering the same trend – in Australia international students comprised 24% of the total in 2005 (http://www.idp.com/).

This mix of students enriches the cohort from a cultural perspective and, of course, contributes not only to university finances – typically, the 'overseas' student fee is higher than that for a home student – but also to the general economy. However, this educational tourism also has its drawbacks because international students may be ill-prepared to study away from home. For an in-depth discussion of this subject, see Carroll and Ryan (2005).

Summary

In this chapter we have reviewed the educational landscape in the UK, to provide a backdrop for understanding the context in which universities operate. Institutions are assessed in terms of excellence in research, teaching and learning. Universities are expected to achieve certain targets in each area and, whether or not they do, both reflects the activity of individual academics and impacts on the way those academics can operate in the future. It is essential therefore that all academics are aware of the drivers and criteria for assessment.

Selected reading

Texts:
Carroll, J. & Ryan, J. (2005). *Teaching International Students: Improving Learning For All*. London; New York: Routledge.
Fry, H., Ketteridge, S. & Marshall, S. (2009). *A Handbook for Teaching and Learning in Higher Education: Enhancing Academic Practice*. New York, London: Routledge.

Educational journals

Active Learning in Higher Education – http://alh.sagepub.com/
Assessment and Evaluation in Higher Education – http://www.tandf.co.uk/journals/titles/02602938.asp
British Educational Research Journal – http://www.tandf.co.uk/journals/carfax/01411926.html
Higher Education – http://www.springer.com/education+%26+language/higher+education/journal/10734
Studies in Higher Education – http://www.tandf.co.uk/journals/carfax/03075079.html

Web resources

1994 Group Universities – http://www.1994group.ac.uk/
Department of Education for England – http://www.education.gov.uk/rsgateway/
Department of Employment and Learning – http://www.delni.gov.uk/
Higher Education Academy – http://www.heacademy.ac.uk/home
Higher Education Funding Council England – http://www.hefce.ac.uk/
Higher Education Statistics Agency – http://www.hesa.ac.uk/
IDP Education – http://www.idp.com/
Million+ Group Universities – http://www.millionplus.ac.uk
National Foundation for Educational Research (NFER) – http://www.nfer.ac.uk/nfer/index.cfm
Quality Assurance Agency – http://www.qaa.ac.uk/
Research Excellence Framework – http://www.hefce.ac.uk/research/ref/
Russell Group Universities – http://www.russellgroup.ac.uk/
University Alliance Group Universities – http://www.university-alliance.ac.uk/
University and Colleges Admissions Service – http://www.ucas.com/

Index